LAMENT FOR AN ARMY

The Decline of Canadian Military Professionalism

Titles in the *Contemporary Affairs* Series

CONTEMPORARY *Affairs* NUMBER 3

LAMENT FOR AN ARMY
The Decline of Canadian Military Professionalism

John A. English

CANADIAN INSTITUT
INSTITUTE OF CANADIEN DES
INTERNATIONAL AFFAIRES
AFFAIRS INTERNATIONALES

CIIA/ICAI

IRWIN
PUBLISHING

Toronto, Canada

Copyright © 1998 by Irwin Publishing. All rights reserved.

Canadian Cataloguing in Publication Data

English, John A. (John Alan)
   Lament for an army : the decline of Canadian military professionalism
(Contemporary affairs series ; no. 3)
Includes bibliographical references and index.
ISBN 0-7725-2520-X

1. Canada. Canadian Armed Forces — Officers — Professional ethics. 2. Command
of troops. 3. Military ethics — Canada. 4. Canada. Canadian Armed Forces —
History.
I. Title. II. Series

UB210.E53 1998   355.3'3041'0971   C97-932633-8

*The Canadian Institute of International Affairs is a national, non-partisan, non-profit
organization with a mandate to promote the informed discussion, debate and analysis
of foreign policy and international affairs from a Canadian perspective. By virtue of its
constitution, the CIIA is precluded from expressing an institutional opinion on these
issues. The views expressed in the Contemporary Affairs series are, therefore, those of
the author alone.*

This work could not have been completed without the generous support of the
Social Sciences and Humanities Research Council of Canada.

Cover Photo: Canapress Photo Service (Ron Frehm)
Design by Sonya V. Thursby / Opus House Incorporated
Typesetting by Opus House Incorporated
Edited by Norma Pettit

Published by
Irwin Publishing
1800 Steeles Avenue West
Concord, ON
L4K 2P3

To my Mother
Amelia Lydia Stuber
of Prussia, Saskatchewan
1913-1997
"And God shall fold her with soft wings."

Contents

Foreword

The Canadian Institute of International Affairs, in cooperation with Irwin Publishing, is very pleased to launch its new Contemporary Affairs series of monographs with John English's examination of Canadian military professionalism. A retired lieutenant-colonel of infantry, the holder of doctorate in history, and the author of internationally hailed books on warfare, John English is the ideal writer to probe this vital subject.

Indeed, no subject has been more in the news in Canada in the last several years. The tragic events in Somalia, as well as the difficulties encountered by the Canadian Forces in the Former Yugoslavia and in Haiti, have thrust the Canadian military into an unaccustomed and uncomfortable public spotlight. The difficulties at National Defence Headquarters and the revelations of the Somalia Inquiry have also called into question the survival of the military verities on which Canadians had come to rely.

Dr. English's views are strong; so is his language. Many readers will agree with him, but some will not. What is important as we draw near the end of this last decade of the twentieth century is that the debate about defence be engaged with Canada's national interest firmly in mind; part of this discussion must involve the type of Canadian Forces we need and the nature of the military professionalism required to achieve them. John English's book is essential reading for this debate.

J. L. Granatstein
Rowell Jackman Resident Fellow
Canadian Institute of International Affairs

Introduction

A *lament*, as explained by Canadian philosopher George Grant in his eloquent *Lament for a Nation: the Defeat of Canadian Nationalism*, is "to cry out at the death or dying of something loved." Yet, as Grant went on to say, lamentation need not be an indulgence in cynicism or despair, for in any lament there is not only pain and regret, but also celebration of passed good. My lament mourns the decline of a Canadian military professionalism that once flowered like a field of poppies within a military that for half a century ensured the honour of Canadian arms. It seeks also to stir the soul of the reader so that serious thought may be given to what laid military professionalism and honour low in the dust and sand of Somalia. The end result of such reflection, it is to be hoped, will be to spark a renaissance within the Canadian armed forces, for the Somalia Affair was as much a symptom as the lowest point of a waning military professionalism.

The concept of *professionalism* has been defined in many ways, but it seems to be generally agreed that without superior knowledge, practical expertise, and a sense of corporate responsibility there can be no professionalism. The true professional is driven primarily by a vocational thirst for knowledge and desire to serve others rather than by hope of financial remuneration. The obligation to perform selfless service to society, even to the point of death if required, distinguishes the military professional from most other professionals. Traditional codes of ethical conduct guiding personal behaviour and leadership have additionally constituted hallmarks of professionalism. Not surprisingly, given the foregoing parameters, the internal ordering and policing of the military profession in peace and war has generally fallen upon the officer corps within most progressive armed forces. No military force can be a good force without an officer corps made up of leaders who have been educated and trained so that they are imbued with broad conceptual knowledge, hands-on expertise, and a deep-seated sense of obligation. Put simply, the officer corps constitutes the head of the body mass of the armed forces made up of the rank and file and their "backbone," the non-commissioned officers (NCOs) who provide the vital link between them and the officer corps. In practice, NCOs remain responsible for the enforcement of unit discipline and the execution of orders given by their officers, to the point that they have been deemed to be indispensable "regimental property," but they cannot by definition be considered professional in the sense of being military "trustees" or keepers of the keys. The thrust of this work will accordingly be directed upon what might not inappropriately be called the "brain" of the armed forces.

As the senior service in historical fact if not in traditional precedence within the Canadian armed forces, the army provides a broad avenue for exploring the state of Canadian military professionalism. The little known and less studied history of the Canadian army also offers more than a few penetrating insights into the traditional reactions of the Canadian people and their governments during crises that resulted in the implementation of military measures. One thing seems clear, that no matter how much their neglect of their armed forces in peacetime, the Canadian people in time of war can be expected to want to field forces larger than those maintained in times of peace. Public opinion in the past has additionally demanded the overseas deployment of Canadian *land* forces, largely because of the widespread perception that only troops on the ground really signify commitment. The problem with military establishments is that they often lose sight of these historical realities and, instead of accepting peacetime public disinterest as an inevitable but dormant condition to be worked round, compromise their professionalism by searching for non-warfighting roles relevant only to gaining them bureaucratic approbation. On reflection, however, they should perhaps be expected to know that the Canadian people, who thrice forced their governments to conscript in the past, might well come knocking again on the door of the military establishment to get a competent officership to lead their sons and daughters into action in defence of Canada.

In the course of these pages, the growth and development of the Canadian army, and most especially its leadership, will be traced with a view to determining what needs to be done in the aftermath of the worst crisis in the history of Canada's military establishment. In contrast to the conscription crises of 1917, 1942, and 1944, which were largely national issues whose echoes reverberated within the halls of the defence department, the Somalia Affair shook the professional military establishment to its very foundations. That the casualty list eventually included a Minister of National Defence, a Deputy Minister, three Chiefs of the Defence Staff (CDS), and the Canadian Airborne Regiment itself further pointed to a worrisome malaise within a Canadian defence establishment that had forgotten too much and learned too little. The army, in particular, had clearly lost its way. To determine the underlying causes of this deviation, one has to look at the history of the Canadian army before Somalia and develop an understanding of the true nature and purpose of armies. Of necessity, this must also include acquiring a feel for how armies actually train and work, for all too often the complexities of their organization and the conduct of their operations have not been fully appreciated even by those retained to know. The secrets of generalship, in short, are not to be found in the corridors of National Defence Headquarters (NDHQ), but in the trials and tribulations of real Canadian commanders who faced the challenge of leading Canadian troops in the ultimate test of battle. The true measure of army leadership still remains performance in the field, and in Somalia Canadian army leadership did not measure up.

List of Abbreviations

AG	Adjutant-General	MRG	Management Review Group
APC	Armoured Personnel Carrier	NATO	North Atlantic Treaty
BEF	British Expeditionary Force		Organization
BCATP	British Commonwealth Air	NCO	Non-Commissioned Officer
	Training Plan	NDHQ	National Defence Headquarters
CAR	Canadian Airborne Regiment	OMFC	Overseas Military Forces of
CASC	Canadian Army Staff College		Canada
CAST	Combined Air-Sea Transportable	PIAT	Projector, Infantry Anti-Tank
CCOS	Canadian Chiefs of Staff	QMG	Quartermaster-General
CDS	Chief of the Defence Staff	RAF	Royal Air Force
CEF	Canadian Expeditionary Force	RCAF	Royal Canadian Air Force
CFHQ	Canadian Forces Headquarters	RCN	Royal Canadian Navy
CFSC	Canadian Forces Staff College	RHLI	Royal Hamilton Light Infantry
CID	Committee of Imperial Defence	RMC	Royal Military College
CJFS	Canadian Joint Force Somalia	SCR	Security Council Resolution
CGS	Chief of the General Staff	UN	United Nations
CLFCSC	Canadian Land Forces Command	US	United States
	and Staff College	UNOSOM	United Nations Operation in
COS	Chief of Staff		Somalia
CRA	Commander, Royal Artillery	UNITAF	Unified Task Force
	(divisional artillery commander)	VC	Victoria Cross
DM	Deputy Minister		
DND	Department of National Defence		
GHQ	General Headquarters		
GOC	General Officer Commanding		
GS	General Service		
GSO 1	General Staff Officer Grade 1		
HD	Home Defence		
MC	Military Cross		
MGO	Master-General of Ordnance		

Chapter One

The Mirror Crack'd in Somalia

In December 1992 by order-in-council the government of Prime Minister Brian Mulroney committed Canadian troops to participate in the US-led United Nations Unified Task Force (UNITAF) charged with establishing by all necessary means a secure environment for UN humanitarian relief operations in civil-war-torn Somalia. The Canadian Joint Force Somalia (CJFS) comprised the support ship HMCS *Preserver*, various air elements, and an 850-man battle group consisting of the Canadian Airborne Regiment (CAR) reinforced by a light armoured squadron of the Royal Canadian Dragoons, and the mortar platoon of the 1st Battalion, The Royal Canadian Regiment. Originally, the CAR had been earmarked to participate in the UN Operation in Somalia (UNOSOM) established by Security Council Resolution (SCR) 751 of 24 April 1992 and expanded by SCR 775 of 28 August 1992. But deteriorating conditions in Somalia, coupled with the realization that attempts to deal with local warlords had reached an impasse, led the UN Secretary-General Boutros Boutros-Ghali to agree to peace enforcement action by UNITAF, which was authorized by SCR 794 of 3 December 1992. Planning nonetheless continued to call for UNITAF to be superseded by a second UN-controlled operation designated UNOSOM II, the purpose of which was to complete the humanitarian relief efforts begun by UNOSOM in September and UNITAF in December. In keeping with this intent, US forces began to withdraw in March 1993 and UNITAF handed over its responsibility to UNOSOM II in May. The mission of the CJFS throughout this period was to "provide, as part of the multi-national Unified Task Force, the secure environment necessary for the distribution of humanitarian supplies in Somalia."

Although many Canadian soldiers performed well in Somalia, the CAR paratroopers who murdered a Somali prisoner on 16 March 1993 committed an atrocity akin to those visited upon Canadian prisoners by SS troops during the Battle of Normandy in 1944 in the Second World War. The difference was that the SS executed Canadian captives in cold blood in real war; Canadian soldiers in Somalia brutally tortured their prisoner to death in a setting in which they never once came under effective hostile fire (there was not one Canadian battle casualty in Somalia). Although victims in both cases were entitled to humane treatment and the full rights and privileges of protected persons under international law, in both cases, officers and NCOs who knew or should have made it their business to know what was going on did nothing to intervene. Some similarity also has been seen between the behaviour of the commander of the 25th SS Panzer Grenadier Regiment *Standartenfuehrer* Kurt Meyer, who incited the men under his command to deny quarter to Allied troops, and the rumours abounding in Somalia that Colonel Serge Labbé, the Canadian national commander in Somalia, had promised a case of champagne to the first soldier who killed a local. Labbé strongly denied such rumours.

To compare the conduct of a Canadian army unit in Somalia with that of an SS formation in Normandy will strike some as overly harsh. Yet, in the heinous matter of killing prisoners there exists but a difference of scale. In two counts out of three, Kurt Meyer was found to be "responsible," by *Canadian* officers who sat in judgement upon him, for the deaths of eighteen Canadian prisoners in the custody and care of the 12th SS Panzer Division. By the same criterion, the heads of senior Canadian commanders in Somalia should most certainly have rolled for the physical mutilation and death of a detainee legally under the protection of the Canadian state.

In the old Canadian army, that is to say the army that existed prior to the unification of the Canadian services in 1968 and briefly in lingering spirit thereafter, the standard or "school" solution to what happened in Somalia would have been to fire both Labbé and the Commanding Officer of the CAR, Lieutenant-Colonel Carol Mathieu, as they were the responsible commanders. Such simple forthright action would have rendered the punishment of subordinate officers and men more palatable, ensured the survival of a reformed Canadian Airborne Regiment, and negated the need for any outside commission of investigation. The new age approach of the Canadian military establishment to employ spin-doctor techniques, on the other hand, served only to exacerbate matters by fuelling charges of cover-up. Like French army officialdom during the infamous Dreyfus Affair of 1897-8[1], senior defence bureaucrats at NDHQ unwisely chose to circle the wagons rather than aggressively seek out the unvarnished truth. Protecting the image of the Canadian Forces, in short, became more important than addressing the reality of what was by any traditional military yardstick a serious command and disciplinary problem. The Deputy Chief of the Defence Staff Admiral Larry Murray even encouraged Labbé to write his privately (and selectively) distributed *Canadians in Somalia—Setting the Record Straight or The Somalia Cover Up*, which extolled the mission's humanitarian efforts while downplaying an obvious atrocity. Other serving soldiers, in stark contrast, remained muzzled by a ministerial gag order. By attempting to conceal what really went on in Somalia through damage-limitation measures rather than dealing with an atrocity that stared it in the face, the Canadian defence establishment seriously undermined its credibility.

That the commanders' heads did not roll, while the heads of private soldiers and NCOs did, left another dark stain upon a Canadian army that once equated command with responsibility. The old Canadian army held that there were no bad soldiers, only bad officers. The rough and ready rule of thumb was to sort out commanders before dealing with their subordinates. Like Napoleon, the General Staff in the old army had also recognized that top commanders had sometimes to be sacked merely because they were unlucky. To defend the unsuccessful or the unlucky, in other words, was often as detrimental to the effective functioning of an army as was defending the indefensible.

Whether Canadian senior commanders were simply unlucky in Somalia remains highly questionable. Besides the obvious torture-murder of a prisoner, Canadian troops may also have carried out an execution-style killing of a wounded Somali downed by small arms fire. The Canadian medical officer who made this allegation in Somalia had to be given an armed escort for fear that he might be "fragged" by paratroopers. Meanwhile, captured petty thieves and infiltrators continued to be improperly bound with ropes in an approximation of shackling contrary to the spirit of the Geneva Convention. Video camera recordings of swastika-tattooed Canadian Airborne thugs drinking beer while brandishing weapons provided further evidence of a serious breakdown in unit discipline and military leadership. For such behaviour to have occurred could only have meant that senior commanders either never inspected or that they simply did not comprehend the practical wisdom of old army prohibitions against such a potentially explosive mix of weapons and alcohol. That none of their subordinate officers or NCOs saw fit to take corrective action furthermore attested to a cavalier approach more reminiscent of swashbucklers than soldiers (and which may have accounted for the death of one paratrooper from the accidental discharge of another's rifle). Sloppy, derisive, and unprofessional orders that recommended shooting intruders "between the flip-flops and loin cloth," ostensibly to reduce the effect of fire, likewise stood at technical variance with General George Patton's sanguinary warfighting advice to "Shoot short. Ricochets make nastier sounds and wounds."

Obviously, it is not necessary to recount the entire litany of things that went wrong in Somalia to conclude that the conduct of the mission reflected poorly on the state of Canadian military professionalism. To be fair, Canadian peacekeepers had beaten locals before and even killed Clito, the legendary Greek-Cypriot immortalized by English writer Laurence Durrell in his book *Bitter Lemons*. But individual perpetrators had always been brought to justice before. Moreover, what happened in Somalia was not the result of a public street fight or a barroom brawl. Rather it was the manifestation of a serious indiscipline *on duty* that allowed a defenceless prisoner to be beaten to death while under the care and protection of a formed unit of the Canadian army. By this not inconsiderable measure, the conduct of a Canadian unit engaged in peace operations was hardly morally superior to that of a SS troop formation barbarized by ideology and real war. That Colonel Labbé barely mentioned the cowardly killing of a prisoner in his narcissistic *Canadians in Somalia* revealed an additionally disturbing amorality at the very highest level of command.

Certain shortcomings that illustrated the depth of the Canadian military establishment's intrinsic amateurism could also be detected in the Somalia mission. At the highest level, the structure of the command relationship was seriously flawed in that it did not provide for adequate force structural checks on the exercise of Canadian operational command. Instead of following the proven General Staff practice of balancing personal strengths and weaknesses in the selection and appointment of commanders and staff, NDHQ senior management named officers of almost identical background to the highest command positions in Somalia. The Canadian national commander, Colonel Labbé, was not only a francophone infantryman and "Van Doo" like Lieutenant-Colonel Mathieu who commanded the Canadian Airborne Regiment, but he was also a former member of the that regiment. To have expected an officer of such similar background to exert as effective a controlling influence as, say, a hard-nosed anglophone colonel of another arm or corps was surely naive. In other words, placing a tank or artillery officer over an infantryman, as was General Staff practice in the old army, would have provided additional balances and checks within the Somalia theatre beyond that of personality alone. In the event, Labbé focussed so excessively upon the non-military aspects of his mission that he neglected to exercise the necessary disciplinary control expected of a Canadian national commander. There is also reason to suspect that he collaborated with Mathieu to downplay the suspicious circumstances surrounding the killing of Somalis. One of his first messages sent allegedly stated that the Somali prisoner received his wounds while resisting capture.

By failing to comprehend and respond with moral courage—with military professionalism—to the pernicious nature of the war crime perpetrated by Canadian soldiers on duty, NDHQ senior management also displayed a notable lack of professional ethics. But, the subsequent elevation of Colonel Labbé to the commandantcy of the Canadian Land Forces Command and Staff College (CLFCSC) constituted by far the most damning indictment of the state of Canadian military professionalism. The CLFCSC was the post-unification successor institution of the Canadian Army Staff College (CASC) first established in Britain during 1940 and later moved to Kingston, Ontario. Like the British Staff College, after which it was modelled, the CASC was the nursery of the General Staff and the single most important educational institution in the army. The Commandant of the CASC reported directly to the Chief of the General Staff who took a personal interest in the training of his staff officers. The CASC not only granted the coveted "psc" (passed staff college) degree to senior captains, but remained the army's centre of operational and doctrinal expertise. Although the CLFCSC constituted a truncated version of the CASC, it continued to conduct professional training considered vital to ground force operations and thereby withstood repeated attempts to shut it down. Grass-roots support for the CLFC-SC within the army at large also meant that the selection of CLFCSC commandants could not be taken lightly. Indeed, as CLFCSC instruction was based on Socratic method and conducted by experienced officers of character, commandants had to be as nearly as possible of stainless reputation.

A pejorative NDHQ view of CLFCSC nonetheless persisted, often reflecting poorly on a top army brass who refused to see the CLFCSC as the last vestige of the old professional Canadian General Staff. The appointment of Labbé showed just what little importance the then CDS General A.J.G.D. de Chastelain attached to the CLFCSC and to the professional training of army officers. Apart from having presided over the most infamous mission in Canadian peacekeeping history, Labbé was not qualified either by intellectual inclination or by training to assume this influential post. He had never been a staff college instructor nor a brigade commander. By traditional criteria, even distinguished service as Canadian national commander in Somalia responsible mainly for administration and discipline could not have offset this lack of experience. Besides, Labbé had originally been slated to command the more junior Royal Military College (RMC) as a "one star" brigadier-general following a successful ticket-punching tour in Somalia. One suspects that it was his desire to command RMC that drove the ambitious Labbé to deny rather than admit any wrongdoing in Somalia. With no stars to steer by other than those in his eyes, he even launched damage-control "counter-offensives" against what he termed "lies [that] fed on lies." Once the taint of Somalia rendered Labbé unsuitable as a role model for RMC cadets, he should never have been placed in authority over still impressionable officers in the more senior and important, though less visible, CLFCSC.

The conclusion seems inescapable that de Chastelain, who along with his Vice-Chief gave the final stamp of approval on the posting and promotion of colonels, used the CLFCSC as a convenient hiding place for Labbé in the hope that the Somalia Affair would eventually blow over. In so doing, he not only denigrated the CLFCSC as an institution, but allowed Labbé to participate directly in the professional training of the army's most promising young officers. The position of commandant also provided Labbé with a captive audience and podium from which to specially plead his own case. The result was that over a two-year period Labbé was able to proactively promote the Somalia expedition as an example of successful Canadian peacekeeping. He was additionally able to ingratiate himself with superiors in the chain of command and forestall criticism from subordinates on whom he was empowered to render annual performance evaluation reports.

If anything, the privileged treatment accorded Labbé clearly demonstrated that the military establishment retained an unshakable faith in the infallibility of its officer selection process. In a supremely revealing act of denial that bordered on outright refusal to see, the commander of the army as late as 1996 brazenly rated the former Canadian national commander in Somalia the best colonel in the army. Such unqualified praise, in fact, highlighted the plight of the army and the Canadian Forces at large. The enemy was within. Left to its own devices, the Canadian military establishment would never have brought the Somalia Affair to light. Admiral John

Anderson, the CDS at the time, chose to *visit* rather than *inspect* the Canadian Joint Task Force in Somalia and therefore found no indications of wrongdoing. According to Labbé, he did not want to get involved in details. The day after the alleged execution-style killing of a Somali, Anderson cautioned Canadian troops not to cause defence minister Kim Campbell embarrassment during her campaign for the prime ministership. For bungling the handling of the Somalia Affair when it first broke in Canada, thanks mainly to a free press and a few honest soldiers, he was subsequently kicked upstairs by the Liberal government to become Canada's ambassador to NATO. In arranging Anderson's replacement, the government also set an unhealthy precedent for the *overt* politicization of the military establishment by recalling his predecessor de Chastelain, who had been named Canada's ambassador to Washington by the previous Conservative government. Since de Chastelain had recommended Canadian participation in Somalia in the first place, he might have been easily persuaded to accept the lead of the Deputy Defence Minister Robert Fowler in hushing matters up. For his pains, Fowler was later made Canadian ambassador to the United Nations. Meanwhile, Labbé not only completed his full tour as commandant of CLFCSC, but was rewarded with an overseas posting to NATO Headquarters in the shadow of Anderson. One has to ask, how ever did the Canadian army get to this state?

NOTES

1. Alfred Dreyfus was a Jewish French army officer who was arrested on 15 October 1894 on suspicion of spying. Despite his protestations of innocence, he was found guilty and was suspended from the army and sentenced to life imprisonment on Devils Island. In 1896 a member of the French General Staff found documents that convinced him of Dreyfus' innocence, but his superiors ordered him to drop the matter. Many people outside the army worked hard to get a new trial for Dreyfus. However, his second trial in 1899 was a mockery because feeling against Jews was so bitter in the army. Many officials felt that the case was closed and that the army's honour was at stake.

Testimony favourable to Dreyfus was barred, and the court again found him guilty. People throughout the world protested the trial and finally, in 1906, the case was reviewed by the highest court in France and Dreyfus was declared innocent.

Chapter Two

Shock Army of the British Empire

The professionalization of the Canadian army dates from the reforms of Sir Frederick William Borden, Liberal Minister of Militia and Defence from 1896 to 1911. Perhaps the best of but two or three distinguished defence ministers in Canadian history, Borden strove mightily to create a militarily autonomous, self-contained citizen army or militia based on part-time service that would be capable of cooperating in the defence of the British Empire on an equal footing with contingents from other countries in the Empire. To ensure that this Canadian militia possessed the organizational structure to maintain an army in the field, he authorized the establishment of a permanent force element of supporting medical, engineer, army service, ordnance, signals, clerk, and pay corps. Although Borden always held that the primary purpose of the permanent force of regular regiments was to instruct the part-time militia or reserve force, he eventually agreed that the permanent force should also be allowed to train as an embryonic professional standing army. Of far greater import, however, was his singular insistence on the creation of a Canadian General Staff.

The organizational institution of the General Staff had been the greatest military innovation of the nineteenth century and reflected a growing need for professional expertise in the execution of field command. As discovered by Napoleon in Russia, there were definite limits to the size of an army, however well drilled and or disciplined, that could be controlled by one person. Military genius alone was no longer sufficient to shore up the generalship of large forces. The solution, first instituted by the Prussians to compensate for lack of military competence among generals who were often appointed for reasons of birth alone, was to introduce the concept of the "general staff with troops," whereby field commanders were provided with advisers capable of offering expert counsel and overseeing the detailed execution of orders. In solving the increasingly complex problems of modern armies related to people, things, and fighting, the staff acted as the lubricant to the chain of command that previously turned solely on councils of war between commanders and subordinates. A General Staff also enabled armies to study war in peacetime and devise practical doctrines for warfighting that would take too long to fashion in actual battle. The establishment of the highly professional Prusso-German General Staff, in fact, represented the triumph of knowledge and function over rank and hierarchy.

It took the humbling experience of the South African War (1899-1902), also known as the Boer War, in which British and Empire soldiers had a difficult time defeating Boer commando and guerrilla forces, to spark serious British military reform. In 1904 the War Office (Reconstitution) Committee headed by Viscount Reginald Esher recommended replacing the traditional post of Commander-in-Chief with an Army Council that would be more compatible with the parliamentary system of government. Arguing that the British Empire was pre-eminently a naval, Indian, and colonial power, the committee further proposed the creation of a civil-military cabinet committee, the Committee of Imperial Defence (CID), to ensure in peace that national effort would be correctly coordinated in war. In order to maintain military focus on war preparation and training for war as opposed to peacetime administration, the committee additionally advocated the institution of an elitist "blue ribbon" General Staff to be recruited mainly from the Staff College, Camberley, which had been in existence since 1857. The hope was that, given a permanent nucleus, the CID could become a "Great General Staff" of Empire providing advice on broad military policy. In the event, the CID received but a small secretariat, while the army General Staff, established on a firm basis 12 September 1906, reflected the organization of the newly created Army Council.

Chaired by the Secretary of State for War, the Army Council comprised a civil member, a financial member, and four military members each responsible for his own staff branch. The first military member, the Chief of the General Staff (CGS), assumed responsibility for staff coordination as well as "G" or operational matters that included intelligence, training, and doctrine. The second, the Adjutant-General (AG), looked after "A" or administrative matters concerning individual soldiers, which included mobilization and army schools. The third, the Quartermaster-General (QMG), took care of "Q" or materiel needs such as supply, quartering, and movement. And the fourth, the Master-General of Ordnance (MGO), was responsible for armament, fortification, and procurement. With the exception of the MGO whose responsibilities were discharged by "Q" branch, staff representation on the Army Council projected downward throughout the field army. Like the corporate yet decentralized Prusso-German prototype, the British staff system comprised central planning "capital" as well as field "troop staff" elements, but it did not constitute a separate caste unified under one chief. The CGS, though *primus inter pares* on the Army Council, headed only one of three autonomous staff branches.

In November 1904 the Canadian government emulated the British by replacing the General Officer commanding the Canadian Militia with an advisory Militia Council. Although this action may have been taken more for political than military reasons, it had the effect of establishing a separate Canadian General Staff system. Presided over by the Minister of Militia and Defence, the Militia Council comprised a civil member (the Deputy Minister), a financial member, and four military members: CGS, AG, QMG, and MGO. From 1903, selected Canadian regular officers also began to attend the Staff College, Camberley. The failure of two candidates to pass entrance examinations for the 1906-7 course and the continued difficulties initially experienced by those who attended Camberley prompted the introduction of a short preparatory staff course at the Royal Military College of Canada. In 1908, a year after Canada had accepted the need for an Imperial General Staff[1], a militia staff course consisting of theoretical and practical portions, each punctuated by examinations, was reinstituted to train members of the reserve militia in general and administrative staff duties. When the Great War broke out in August 1914 there were 124 successful graduates of this course, which, though inadequate by Camberley standards, served to alleviate a shortage of staff officers within the Canadian General Staff. A total of twelve Canadians had graduated from the two-year course at Camberley and three were undergoing training.

Yet, as pointed out by Stephen Harris in *Canadian Brass*, the amateur ethos of the Canadian militia ran counter to the concept of professionalism. The myth that citizen soldiers had won the War of 1812 against the Americans encouraged the perception that a volunteer militia comprised of part-time soldiers was the only army that Canada needed. Besides, defence was an imperial problem to be financed and addressed by the British if things went wrong. Appointments and promotions within the Canadian militia had also been an integral part of the government's patronage system. Even Borden, who gradually came to accept that patronage was no basis for the promotion and appointment of regulars, continued to believe that politics and military service could be mixed in the part-time militia. By the time Sam Hughes succeeded Borden as Minister of Militia and Defence in 1911, the militia with a strength approaching 60,000 outnumbered the regular force twenty to one. Hughes drew no distinction between regular forces and the militia, and in a throwback to earlier times dispensed both patronage and myth with the fervour of a zealot. A long-time volunteer militia officer, Hughes saw little merit in maintaining a small standing army. He despised professional soldiers and cited the lacklustre performance of British regulars in the South African War as proof positive of the inherent superiority of the Canadian citizen soldier. As if to illustrate his case, he personally promoted himself from colonel to major-general.

Not surprisingly, Hughes had no confidence in the Canadian General Staff, which as early as 1911 had prepared a contingency plan for the dispatch of an overseas expeditionary force of one division and a cavalry brigade. The precedent for sending troops abroad had, of course, been set during the South African War when Canadian public opinion forced the government of Sir Wilfrid Laurier to authorize the dispatch of a volunteer contingent to serve with the British Army. In the South African War, regular force regiments were among the first to be committed. However, the 1911 mobilization scheme called for fielding composite units drawn from various militia regiments within national regions according to population and using regulars to train them. This scheme not only avoided the difficulty of having to choose one reserve militia unit over another, which invariably triggered regimental infighting, but also ensured the more even distribution of potential casualties nationally. To minimize the confusion usually associated with raising new units, decentralized command and control measures designed to prevent the premature concentration of the entire expeditionary force in Camp Petawawa were prescribed. On the outbreak of the Great War, however, Hughes scrapped both this mobilization plan and a 1914 variant that recommended basing the infantry division on existing reserve militia regiments rather than composite units. He instead ordered the construction of a new army camp at Valcartier, Quebec (largely for venal reasons related to awarding contracts to friends, for Petawawa was a superior functioning training area) and directly summoned over a hundred militia battalions to immediately concentrate there with as many men as possible so that composite units for divisions could be selected on the spot. The result of this arbitrary ministerial intervention was mass confusion and internecine regimental conflict on a grand scale.

On observing the disembarkation of the Canadian Expeditionary Force (CEF) in Britain in October 1914, the future British military reformer J.F.C. Fuller caustically remarked that Canadian soldiers would be good enough after six months training if "the officers could be all shot." That there was a certain ring of truth to Fuller's extreme statement could hardly be denied as Hughes had blatantly resorted to patronage in the appointment of officers at Valcartier. In giving command of the 1st Canadian Infantry Division to a Britisher, Major-General E.A. Alderson, Canadian authorities other than Hughes also acknowledged the military shortcomings of Canadian officers. Ironically, Alderson was later fired as commander of the Canadian Corps for the incompetence of his subordinate commanders in the battle of the St. Eloi craters in Flanders in April 1916. Here disorientated troops of the 2nd Canadian Infantry Division captured the wrong objectives, thinking and reporting that they were the right ones. Unfortunately, neither their brigade commander Brigadier-General H.D.B. Ketchen, nor their divisional commander Major-General R.E.W. Turner endeavoured to find out the real situation though ample intelligence and reconnaissance means existed. In the case of Turner, this was the second time that he had displayed incompetence in command of a field formation. However, he was a friend of Hughes and a Victoria Cross winner to boot. Largely because of Canadian behind-the-scenes political manoeuvrings, it was ultimately the corps commander rather than the divisional commander who was held responsible for the St. Eloi fiasco.

15

When Alderson was dismissed in 1916 his replacement was Lieutenant-General Sir Julian Byng who went on to lead the Canadian Corps in its most celebrated if not most brilliant victory at Vimy Ridge in April 1917. In November 1916 Hughes was also dismissed as defence minister thereby paving the way for the increased professionalization of a Canadian Corps that had long come to reject Hughes' nationalistic amateurish meddling. Indeed, it was Canadian Corps reaction against ramshackle training, reinforcement, and personnel selection policies and procedures that laid the foundation for the development of a truly professional Canadian army. The great majority of Canadian senior officers, regular and reserve, had come to appreciate the value of a non-partisan approach to the conduct of war. To a man, they had opposed Hughes' attempt to replace their British staff officers with Canadians who had no formal staff training. Since established criteria existed for measuring combat performance, the Canadian Corps had also sought more authority and independence in the selection and appointment of military leaders. It was not until 1917, however, that the merit principle of promotion came generally to be applied, largely at the insistence of the newly appointed Minister of the Overseas Military Forces of Canada (OMFC) Sir George Perley, based in London. By this time, of course, Canadian militia soldiers and civilians alike had come to realize that any military system incapable of demonstrating expertise and knowledge would pay a higher price in casualties.

Under Byng and his less popular but more professional Canadian successor, Lieutenant-General Sir Arthur Currie, the Canadian Corps became one of the most formidable fighting formations on the Western Front.[2] In a war of high intensity that left little room for strategic manoeuvre, tactical innovation effected through meticulous staff work was critical to reducing friction. Detailed planning, thorough preparation, and exhaustive training, more than any innate Canadian talent for war, characterized corps operations. If there was any Canadian "way of war," it merely reflected the refinement of British warfighting methods. Much of the operational effectiveness of the Canadian Corps also sprang from the superior skill of attached British staff officers, three of whom rose to become Chief of the Imperial General Staff. Currie himself leaned heavily on his British Brigadier-General, General Staff (principal operations staff officer), and on his British Deputy Adjutant and Quartermaster-General (principal logistics staff officer), adamantly refusing to replace them with Canadians and ultimately singling out the former, N.W. Webber, for particular personal praise. At the divisional level, the first Canadian senior operations staff officer or General Staff Officer Grade 1 (GSO 1) was not appointed until November 1917, and, by Armistice in 1918, one of Canada's four fighting divisions still had a British GSO 1. By 1918, moreover, Canadian divisions had ten more staff officers on their headquarters establishments than their British equivalents.

The Canadian Corps additionally possessed the inestimable advantage of having permanently allocated divisions. This enabled it to develop a military cohesion and operational capacity that can only ever result from soldiers constantly working and living together. British Corps did not have this luxury as the division remained the basic fighting formation permanently structured with minimal artillery and supporting services. To ensure the optimal use of these latter resources, divisions were usually grouped under various corps headquarters, which were in turn assigned appropriate weights of artillery and other supporting troops to carry out specific tasks. This task-orientation of corps also meant that it made good military sense to alter their divisional composition as the operational situation demanded. In the case of the Canadian Corps, however, Currie was able to keep his divisions out of the line during the 1918 German March offensive because he insisted that the Corps fight as a national entity. Currie also refused to triangularize Canadian divisions (by reducing them from four to three brigades each as had the British and Australians), even though doing so would have resulted in his promotion to command an "army" of five divisions in two corps. Canadian divisional establishments thus continued to comprise twelve battalions, each additionally increased by 100 men, as opposed to nine smaller battalions in British and Australian divisions. By 1918 the Canadian division also incorporated a complete machine-gun battalion of 1558 all ranks and 96 Vickers heavy machine guns (compared to the British division's 64). The result was that the Canadian Corps was a far more powerful formation than its nomenclature indicated.

Without question, the Canadian Corps was a cadillac formation. Always kept up to strength, it was never betrayed or let down by the Canadian government after the departure of Hughes. In fact, the close bond between the government of Sir Robert Borden and the CEF precipitated the conscription crisis of 1917. As is too often forgotten when considering the 1917 conscription crisis, it was really only a miracle that the Allies did not lose the war in that year. The year 1917 saw the collapse of Russia, the near collapse of Italy after its defeat in the Battle of Caporetto, widespread mutiny within the French army, the height of unrestricted submarine warfare, and the agony of the Battle of Passchendaele in Flanders fields. Canada's victory in April at Vimy Ridge, a strategically located hill near the French town of Arras, provided one of a few rays of hope.

Faced with serious manpower shortages, the Allies began planning on the basis of the war lasting into 1920. Having gained a strategic voice in the Imperial War Cabinet earlier in March 1917, Borden was well aware of just how bleak the situation was and, accordingly, decided to impose conscription. Ultimately, of the more than nine million armed forces personnel fielded by the British Empire, over a million, mainly soldiers, came from the Dominions and over another million from India. But whereas CEF soldiers had been British imperials recruited under the Army Act before 1916, they were now members of the Canadian militia serving overseas. In 1918 Perley's successor, Sir Edward Kemp, also established a Canadian staff section at British General Headquarters (GHQ) in France, thus ensuring a purely Canadian chain of command running from the Canadian Corps through GHQ and OMFC back to Ottawa.

As demonstrated by Shane Schreiber in his *Shock Army of the British Empire*, the shining hour of the Canadian Corps came during the Hundred Days of victory from 8 August to 11 November 1918. No other Canadian Corps battles, not even Vimy Ridge, ever came close to having as decisive an impact on the outcome of the war as those of the Hundred Days campaign. Time and again, the Canadian Corps was used to crack some of the toughest and most vital points of the German defence, thereby creating the conditions and opportunities that allowed the Allied armies to drive the German military machine to the point of collapse. From the surprise attack at Amiens in France on 8 August 1918, "the black day of the German army," the field forces of the British Empire, using techniques later refined by the German army in *Blitzkrieg* at the beginning of the Second World War, advanced relentlessly over some 100 kilometres to break the back of the German army on the Western Front. To an extent greater perhaps than any other fighting formation, the Canadian Corps spearheaded this last Allied offensive of the war. In terms of planning, preparation, and execution, the offensive was thoroughly modern in nature and involved the use of tanks, indirect fire, tactical air support, chemical weaponry, electronic deception, and command, control, and intelligence systems. In a series of combined arms actions from the breaking of the Drocourt-Queant switch in the German Hindenburg Line in early September to the capture of Mons in Belgium on 11 November, the Kaiser's army was sent reeling back, to be saved from ultimate destruction only by the eleventh-hour Armistice.

That this victory was not easily accomplished was abundantly evident, for the ferocious fighting of the Hundred Days accounted for roughly twenty percent of all Canadian casualties during the entire war. Never before or since have Canadian troops played such a crucial and decisive role in land operations. The exemplary fighting performance of the Canadian Corps was enough to persuade Canada's foremost military historian, the late Colonel C.P. Stacey, that the "creation of the Canadian Corps was the greatest thing that Canada had ever done." The Second World War, he noted with some irony, was but the second-greatest event in Canadian history. In terms of lives lost the price had been enormous, and by such a yardstick the Great War remains Canada's greatest war. The conflict cost the Dominion over 60,000 dead, more of whom fell in battle than their 48,000 American comrades killed in action. Indeed, the Canadian death toll in this far-distant war exceeded that of the United States in Vietnam; but whereas the latter drew its soldiers from a population base of more than 200 million, the Dominion of Canada in the summer of 1914 boasted fewer than eight million souls. Proportionally, although rarely recognized, Canadian Great War battlefield losses equated to the 600,000 soldiers of both sides who perished in the American Civil War. All told, Canada suffered 212,688 combat casualties in its Great Patriotic War. Yet most of its citizens, shamefully, know less about this world-shaking conflict that laid the foundations of Canadian nationhood[3] than they do about the internal struggle that took place south of the border.

From an operational perspective, the Great War still lies heavily on the military landscape, shrouded in the mists of history and myth that becloud the true dimensions of its importance. The forlorn images of the battles of the Somme and Passchendaele continue to overshadow the fire and movement victories of 1918. Yet, the more one studies later wars and military operations, the more one is inexorably drawn back to this watershed in warfighting, which witnessed unparalleled technological change and saw more soldiers perish by fire than by disease. How the Canadian Corps managed to turn a tactical victory into a continuous string of consecutive successes in a sustained campaign during a war of high intensity still demands attention. The relative ease with which Canadians chalked up tactical successes near the end, for example, may have obscured the fact that Currie always employed an all-arms approach incorporating surprise and movement where possible. Modern soldiers, in short, could benefit greatly from the practical lessons to be gleaned from an examination of how the Canadian Corps, which was as large as a small British field army, adjusted to new battlefield realities by adopting and generating highly innovative fighting approaches to war. Unfortunately, failure to produce any written detailed analysis of the corporate knowledge that went into making the Canadian Corps one of the world's most successful fighting formations inhibited the future intellectual growth of the Canadian army. It was left to the German General Staff to study the lessons of the Battle of Amiens and the 100 days that followed.

Four years of trial by fire had nonetheless seen fighting ability and knowledge displace political patronage as a basis for selecting leaders and commanders. The Canadian Corps had of necessity become the most militarily professional ever of Canadian forces that previously had tended to be highly politicized. On repatriation, consequently, there developed a strong movement to have the CEF serve as the foundation of a restructured land force. Better that a dozen old militia regiments should go to the wall, one veteran commander declared, than any CEF units should be lost. A committee headed by septuagenarian Major-General William Otter was accordingly charged in 1919 to make recommendations as to the incorporation of CEF units into the militia. This proved to be a rather complicated, tedious, and controversial undertaking. The committee focussed on parcelling out coveted battle honours won by CEF numbered battalions among those militia units that had furnished the battalions with wartime drafts of soldiers. The original idea of fusing the CEF and militia to ensure the reform as well as reorganization of the latter was soon lost sight of in the process. The lessons of the Great War with respect to the operations of large armies, the role of machine guns and tanks, plus the need for mechanization, were hardly discussed. Little concern was expressed about the purpose and object of military forces or their future role in the defence of Canada. Despite various public calls for a warfighting force rather than a social organization that would have to be jettisoned when storm clouds appeared, the Otter Committee accomplished little more through its deliberations than a return to the prewar situation wherein the militia, as the Canadian army continued to be called, was less a war machine than a social edifice.

The politics of service survival and bureaucratic battles for budget slices also diverted the Canadian militia from its professional focus. In keeping with a policy of retrenchment, the Mackenzie King government from 1922 placed all Canadian services under a single Department of National Defence (DND). This move, strongly supported by the militia, precipitated an interservice struggle of disgraceful proportions. In what was clearly a power play, the CGS Major-General J.H. MacBrien managed to get himself appointed Chief of Staff (COS) with broad powers of coordination over air and naval plans and programs. Subsequent resistance by the Royal Canadian Navy so consumed the COS that matters pertaining to militia preparedness and defence planning were neglected. Indeed, from 1924 to 1927 effective leadership of any kind was lacking within the land forces. As one student of the period put it, the Canadian militia really had no idea what to do with itself other than to keep itself going. On MacBrien's resignation in 1927, the position of COS was allowed to lapse and that of Chief of the Naval Services created. When Major-General A.G.L. McNaughton became CGS in January 1929, he strove unsuccessfully for the reinstitution of the COS position and worked incessantly for a reduction in naval funding. In McNaughton's view, air power had rendered traditional concepts of sea power obsolete, especially in the realm of coastal defence.

Under General McNaughton, who served as CGS from 1929 to 1935, the army focussed mainly upon non-operational peripheral activities that included aid of the civil power call-outs in Nova Scotia, Quebec and Ontario during 1922-3 and 1932-3. At McNaughton's personal instigation, the army between 1932 and 1936 additionally ran relief camps called "Royal Twenty Centres" for unemployed single men. His extensive interest in furthering northern communications and aerial mapping and charting, among other scientific pursuits, also largely contributed to the almost total neglect of training commanders and soldiers for war. To McNaughton, who even revised campaign speeches for the Prime Minister, military knowledge was mainly a matter of technical proficiency that any scientifically educated person could master probably better than a regular officer. Essentially he denied the existence of a profession of arms akin to an academic discipline that called for the detailed and concentrated study of what Carl von Clausewitz termed "fighting proper." These views, scarcely unique to McNaughton, proved attractive and popular to civilian and academic authorities who, in turn, showered honours upon him. When McNaughton left office in 1935 for a $15,000 a year secondment as President of the National Research Council he wrote a famous memorandum describing the deficiencies and sorry state of the Canadian forces, for which several of his former military comrades held him directly accountable.

In effect, McNaughton left the militia in a worse state than it was when he assumed the appointment of CGS. Unlike the circumscribed German *Reichsheer* under Hans von Seeckt, which stressed theoretical as well as practical training for war, the Canadian militia under McNaughton catered to politicians who, as blind as their electorates, could not envision another conflict. In actively seeking and assuming politically attractive non-military roles, ostensibly to ensure the survival of the militia as a fighting force, McNaughton virtually guaranteed the opposite: the continued erosion of whatever operational capability remained. Keeping alive the art of warfighting, especially against a first-rate enemy, was not accorded a high priority.

Notwithstanding the tyranny of Canadian geography, equipment shortages, and lack of communications, much more could have been done in the area of war games, tactical exercises without troops, and skeleton exercises had the will been present. Indeed, certain dedicated officers through *individual* inquisitiveness and diligence demonstrated just what a similarly focussed militia could have accomplished *institutionally*. The tactical debate conducted in the pages of the *Canadian Defence Quarterly* between Lieutenant-Colonel E.L.M. Burns and Captain G.G. Simonds, both future corps commanders[4], stands as a case in point. The doctrinal article produced by Simonds on the "Attack" remains another. Various non-regulars like Brigadier Stanley Todd, who rose to command 2 Canadian Corps artillery, also took it upon themselves to learn all that they could about their particular area of military expertise, so that when war came they would actually *know* something about warfighting. But such focus was neither all that prevalent nor actively promoted by the military establishment itself. For the most part, individual militia officers, both regular and reserve, were left to become professional on their own.

Clearly, electoral blindness that translated into parsimonious defence budgets and political neglect cannot be blamed entirely for the dissipation of the heightened sense of military professionalism that had become the hallmark of the Canadian Corps during the Great War. Neither is it altogether fair to hold peacetime societies more accountable than their soldiers for failing to remember that the chief purpose of any army, in the final analysis, is to fight its country's wars and fight them well. The Canadian High Command was paid handsomely to hold the peacetime army in trust for the Canadian people. In 1932 the CGS received the hefty sum of $10,000 per year, while doctors, dentists, and lawyers around the same time earned but $1500 per annum, schoolteachers $400, and domestics less than $300. Within the defence department in 1932 no fewer than 139 people received between $4000 and $4900 per year, thirty others between $5000 and $5900, and a further fifteen between $6000 and $6900. The pay of regular force brigadiers during this same period ranged between $5000 and $6500 yearly and majors received between $3300 to $4000. People fortunate enough to be on fixed incomes after 1929, which year saw 60% of working men and 82% of working women living on less than $1000 a year, also found that their money could buy much more. Canadian taxpayers thus had every right, despite their disinterest in military affairs, to expect a good return on their money from those they retained as an insurance policy to look after Canada's long-term defence interests.

NOTES

1. Although an Imperial General Staff did eventually come into being in 1909, it was imperial in name only and remained mainly a British staff to which Canadian (and Australian) "sections" were attached.

2. Canadians not only proved adept at mounting raids and massing machine-gun fire, but also developed a matchless corps counterbattery organization that wreaked havoc on German artillery. From all accounts, Canadians were also better disciplined than Australians, though the trench police system of clearing bays and

traverses seemed guaranteed to ensure that most troops went over the top following gallant officers carrying swagger sticks (if the trench police refused to do their duty, they were simply relieved and sent over the top themselves. As it was always safer not to go, this system was nigh on foolproof). Unlike the Australians who refused to sanction the death penalty for their soldiers after the execution of "Breaker" Morant (for reprisal killings) in the South African War, the Canadians also allowed 25 of their number to be executed by firing squad (out of roughly 395 for the Empire, which meant that counting those that were eventually commuted, one death sentence was read up and down the line every week *pour encourager les braves*). As Canadians did not have the straggler problem of the "Diggers," Australian generals perpetually clamoured for the institution of the death penalty.

3. Ten days after the Canadians attacked at Vimy, the British in a resolution of the Imperial War Conference formally recognized Canada as an autonomous nation within an imperial Commonwealth.

4. The highly intelligent Burns ultimately lost command of 1 Canadian Corps in Italy largely because of personality. Called "Smiling Sunray" or "Laughing Boy" by his troops, because he seldom did, Burns never achieved a cordial working relationship with either his British superiors or his Canadian subordinates, particularly Major-Generals Bert Hoffmeister and Chris Vokes. Simonds, in contrast, went on to command 2 Canadian Corps with great distinction.

Chapter Three

Best Little Army in the World

During the Second World War, unfortunately, Canada had to rediscover the hard way just what armies and military professionalism were all about. Although the British had begun to rearm in 1935 and introduced a comprehensive rearmament program the following year, Canada commenced limited rearmament only in 1937. At the Imperial Conference that year King also supported the appeasement policy of British Prime Minister Neville Chamberlain toward the Axis powers and dodged all military commitments. He even refused to cooperate in military planning with the British and forbade Canadian senior officers from discussing measures for the defence of the Crown colony of Newfoundland. In 1938 he rejected British overtures for a large joint air training scheme. As late as April 1939, only two months after Britain had agreed to commit ground forces to the direct defence of France, the King government established a series of defence service priorities reminiscent of those adopted by the Chamberlain government in 1937. The Royal Canadian Air Force (RCAF) was accorded primacy and the Royal Canadian Navy (RCN) second priority. The army of 4000 regulars and 51,000 reservists was left pretty much the Cinderella service. By 1939-40, RCAF appropriations constituted nearly half the total allocation for all three services and exceeded those of the army by eight million dollars. That King had himself insisted upon these priorities is highly likely, since he had asserted in March 1939 that "the days of great expeditionary forces of infantry crossing the oceans are not likely to recur."

As if to reinforce this point, on 24 August 1939 King's isolationist-nationalist Secretary of State for External Affairs, O.D. Skelton, produced a paper entitled "Canadian War Policy" that advocated giving primacy to the defence of Canada. The possibility of aiding Newfoundland and the West Indies also lay, in his view, within Canadian capacity. Skelton further recommended that if any military action was to be taken overseas, it should be in the form of air service rather than military contingents. The announcement of an immediate and intensified program of building planes and training men for air service and for a Canadian air force operating in France, he added, would be effective both from the standpoint of military value and the consolidation of public opinion. It was in the economic field, however, that Skelton envisioned Canada making its greatest national effort. Concentrating on the provision of munitions, raw materials and foodstuffs, he argued, would underpin the most effective contribution Canada could make to its allies and be consistent with Canadian interests. Skelton's proposal amounted to a blueprint for a "limited liability" war, which discredited policy the government of Chamberlain had abandoned along with appeasement six months earlier.

When King read Skelton's paper to the cabinet, apparently without consulting the Canadian Chiefs of Staff (CCOS), it met with general approval. On 1 September the cabinet also considered a CCOS paper entitled "Canada's National Effort (Armed Forces) in the Early Stages of a Major War." Noting that Britain intended to dispatch an expeditionary force to France, this submission advocated raising a Canadian army corps of two divisions and ancilliary troops for overseas service. On 5 September the Cabinet Defence Committee, with King in the chair, informed the CCOS that pending decision by Parliament the government was only prepared to adopt measures for the defence of Canada. King also reportedly voiced his displeasure with their submission, pointing to its dissimilarity with Defence Scheme Number 3 that, in response to political pressure, had been amended in 1937 to provide for a "Mobile Force" of one cavalry (until 1939) and two infantry divisions primarily for the direct defence of Canadian territory (although recognizing the contingency that this force could and might be deployed overseas). Meanwhile, certain highly significant measures had already been taken. On 25 August Royal Navy warships received authorization to use Halifax harbour and after Germany attacked Poland on 1 September the government ordered mobilization. Two days later, when Britain declared war on the Third Reich, a secret order approved by the cabinet ordered Canadian coastal commanders to take all necessary defence measures that would be required in a state of war. For all practical purposes Canada was at war before its formal declaration on 10 September 1939.

On 3 September King also telegraphed Chamberlain asking how the Dominion might possibly provide assistance. Three days later the British government requested naval and air support and made a tactful plea for the immediate dispatch of "a small Canadian [fighting] unit which would take its place along side the United Kingdom troops" and technical units (signals, engineers, ordnance, medical and transportation) for attachment to British formations. On 15 September the cabinet appointed a sub-committee chaired by J.L. Ralston, Minister of Finance, to draft a war program for the nation. As announced on 19 September, the program gave priority to the provision of supplies and financial aid. Canada also offered to provide and pay for technical units, providing that Britain agreed to absorb all costs associated with them while they were not under Canadian command. Although the last item in the war program referred to the organization and training of an expeditionary force, the CCOS had already been instructed on 16 September to dispatch the 1st Canadian Infantry Division overseas to fight as part of a British Expeditionary Force (BEF). The ferocity of the Nazi assault on Poland, which the Russians joined on 17 September, the day German armoured pincers closed near Brest-Litovsk east of Warsaw, had clearly added to the urgency of the situation. It had also raised the spectre of overseas conscription within King's cabinet.

The sorely pressed British government, which had initially requested Canada to assist in the individual training of air crew, including 2000 pilots per year, now proposed a far grander scheme most probably based on earlier Air Ministry war planning. On 26 September Chamberlain personally appealed to King asking Canada to participate in and be host country to the British Commonwealth Air Training Plan (BCATP) initially designed to qualify annually 30,000 British, Canadian, Australian, and New Zealand air crew for service with the Royal Air Force (RAF). To King, the proposal seemed a godsend, for it looked as though most of Canada's war effort could be concentrated on training at home. Apparently convinced of the military merits of air power by British Prime Minister Stanley Baldwin, who in 1932 had declared that "the bomber will always get through," King likewise believed that an air commitment would involve fewer casualties and hence lessen any requirement for conscription. Though expressing regret that the British proposal had not been made earlier, so that Canada could have framed her war effort along these lines instead of having to field expeditionary forces, he cabled his approval to Chamberlain on 28 September. During negotiations, however, King refused to pay the salaries of RCAF air crew overseas or the cost of squadron ground crew maintenance. He also badgered the reluctant British to acknowledge publicly that Canadian participation in the BCATP "would provide for more effective assistance toward ultimate victory than any other form of military cooperation which Canada...[could] give." As the Canadian General Staff foresaw better than King and the elitist Skelton, however, the great majority of the Canadian people were eventually to demand the commitment of a national army to action overseas.

From a purely military perspective, the Second World War further confirmed that the decisive role of armies had not diminished. The Western Front had merely moved east where huge armies clashed in close and deadly combat. After the cataclysmic fall of France in June 1940, Canada became Britain's senior ally and the Canadian army overseas swelled beyond division, beyond corps, to field army size of several corps. On 11 August 1941 the Canadian General Staff estimated that there existed sufficient manpower for Canada to field an overseas "army " of two corps, each of two infantry and one armoured divisions, for a war period of over six years. By the end of the war, in fact, the First Canadian Army comprised two national corps composed of three infantry divisions (each of three infantry brigades), two armoured divisions (each of an armoured and infantry brigade), two independent armoured brigades, two "army groups" (brigades) of artillery, and ancillary troops. Total active strength peaked at 495,804 all ranks compared to 215,200 in the RCAF and 92,441 in the RCN. By the end of 1943 the Canadian Army overseas reached its greatest size as a force of over a quarter of a million men. Despite its national designation, however, the First Canadian Army was never able to operate entirely as a Canadian field force, for it required the permanent commitment of upwards of 9000 men per division from British resources to complete its rearward support and logistic services. As the Canadian Army possessed no heavy artillery, this too had to be provided by the British.

By ministerial agreement, Canadian army organization, doctrine, and training generally conformed to that of the British Army, with which Canadian troops acted "in combination" under terms of the Visiting Forces Act of 1933. In short order, both armies were integrated so closely that one Canadian overseas training school recorded, "there were not two armies, British and Canadian, but one." The formation structure of the Canadian army like the British reflected a regimental system of units, with the difference that the composition of the Canadian expeditionary force divisions had been predetermined so as to give proportional representation to the major territorial regions of Canada. Thus, within the 1st Canadian Infantry Division, units from Ontario made up the 1st Brigade, units from the West the 2nd, and units from Quebec and the Maritimes the 3rd. Similarly, in the 2nd Division, Ontario units made up the 4th Brigade, Quebec units the 5th, and Western units the 6th. The French-speaking complexion of the 5th Brigade was soon compromised, however, when the Quebec-based Royal 22nd Regiment joined the overseas bound 1st Division to ensure that there was francophone regular force representation in the first contingent. The dispatch of Les Fusiliers Mont-Royal to Iceland and their eventual replacement by the Calgary Highlanders later prompted a recommendation from the francophone commander of the 5th Brigade that, owing to a critical shortage of qualified staff officers and commanders, it cease to be a French-speaking formation. Lumping French-speaking units together in one formation was also thought to be unwise given the potential political difficulties that might arise from a heavy incidence of casualties. Indeed, the reinforcement of French-Canadian battalions became an acute problem in 1944 as only four in every hundred French Canadians volunteered for military service compared to ten in every hundred English Canadians.

Since the Militia Act stipulated that no soldier could be compelled to serve continuously in the field for more than a year, volunteers were encouraged to enlist for the duration of the war on general service (GS) for overseas duty in the newly created Canadian Active Service Force. Once inducted, recruits underwent eight weeks common training at Basic Training Centres, after which they passed on to Advanced Training Centres where they received an additional two to five months specialist training. From October 1943 the Number 1 Training Brigade Group, established at Debert, Nova Scotia, also gave reinforcements four weeks of individual and collective training before proceeding overseas. However, after Allied forces were rescued from the shores of Dunkirk in France where they had been trapped by German troops, the National Resources Mobilization Act of 21 June 1940 imposed conscription for home defence (HD) duty. The act, in effect, created two classes of soldiers within the Canadian army. Though both groups trained together from March 1941, differences in status invited comparison and criticism that inevitably translated into outright antagonism. Pressuring HD recruits to go GS exacerbated this military schism as some HD men were allowed to earn six to seven dollars over and above their $1.25 daily soldier's pay for "occupational leave" work on farms and war construction projects. The situation was further aggravated by the fact that GS NCOs proceeding overseas usually had to give up their rank, which other HD men often assumed. This naturally increased tensions as GS soldiers in Canada intensely disliked serving under HD NCOs who, in turn, experienced great difficulty in dealing with GS subordinates. By August 1944 an Adjutant-General Branch survey concluded that it was more difficult to recruit volunteers within the army than without.

Meanwhile, largely because the Canadian army overseas was an untrained citizen force of limited corporate military proficiency, a Canadian Training School was established in Britain in August 1941. Fortunately, the 1st Canadian Infantry Division had more than three years in which to address training shortcomings before being committed to Sicily in July 1943. The less fortunate 2nd Canadian

Infantry Division, which arrived in October 1940 via occupation duties in Iceland, spent little more than twenty-two months in Britain before being decimated on the beaches of Dieppe in August 1942. Thereafter the division enjoyed an equivalent period in which to recover for battle in Normandy. The 1st Canadian Tank Brigade (later the 1st Canadian Armoured Brigade), which accompanied the 1st Canadian Infantry Division to Sicily, arrived in Britain in June 1941. The 3rd Canadian Infantry Division sent over in the autumn of 1941 had more than two and a half years in which to prepare. The 5th Canadian Armoured Division, complete by November 1941, trained for roughly eighteen months in England before departure for the Italian theatre, while the 4th Canadian Armoured Division, which followed in the summer of 1942, spent a similar period in Britain prior to taking part in the Normandy campaign. The 2nd Canadian Army Tank Brigade arrived overseas in June 1943, only ten months before being committed to battle in France.

During these years, the magnitude of the training problem also increased, for it was one thing to train individual soldiers and units, but quite another to train all-arms formations from brigade upwards, which required more intellectual than physical exertions on the part of top army leaders. When German panzers burst through the French front at Sedan in May 1940, the 1st Canadian Infantry Division was still conducting training in trench warfare. Shortly before, the division's General Officer Commanding (GOC), Major-General George Pearkes, VC, had personally demonstrated various methods of crawling with weapons on patrol to a Sandhurst staff course. He later complained that there was too much of a "militia camp" attitude toward training in general.

Until December 1943 all Canadian army elements overseas were commanded by McNaughton, then a political appointee of Mackenzie King. King had personally called McNaughton back to lead Canada's volunteer army overseas. Hailed as a "soldier-scientist," McNaughton had also been built up into a larger than life national figure through a major publicity campaign. In the darkest days of defeat some reporters even went so far as to portray him as a possible leader of an Allied invasion of Europe. Yet, while McNaughton had performed brilliantly as an artillery counter-battery officer in the Great War and was blessed with a compelling personality, his forte was definitely not conducting large formation field operations. According to one observer, he was forever having "attacks of the gadgets" and rushing about in fits of unbounded enthusiasm, at one point even seriously considering the possibility of retaking the Maginot Line in France by employing Canadian Tunnelling Company diamond drillers in counter-mining operations. In preferring to concentrate upon national command and administrative matters, while dabbling in virtually everything else, McNaughton ensured that Canadian army training suffered less from equipment shortages than from a general lack of professional focus. Collective and combined arms training, especially movement and fire coordination, definitely did not receive the attention they merited. In June 1940 he nonetheless signalled Mackenzie King that he now considered his Canadian force, often referred to, tellingly, as "McNaughton's Flying Circus," to be battle worthy. By September he ventured that further intensive training of this already highly trained force would make them stale. He accordingly introduced educational training in academic subjects that would benefit soldiers upon the cessation of hostilities.

The reality was that Canadian formations had still not mastered the profession of arms, especially the intricacies of road movement and getting all fighting elements to work together. In Exercise "Fox," a training exercise held in Britain in February 1941, the 1st Division tied itself up in so many knots that it shook the complacency of everyone from private up. A major problem, quite correctly identified by British General B.L. Montgomery, was that Canadian generals did not actually know how to *conduct* training efficiently, much less have a grasp of what he termed the "stage management" of battle. Indeed, little attention was paid to the training of brigadiers and above. Commanders down even to company level were also permitted to act as exercise "directors" instead of being tested themselves. Few had any idea how to run challenging and stimulating skeleton exercises that would have enhanced the value of collective training, which all too often kept soldiers idle while untrained and unrehearsed officers were put through their paces. The result was that much training never seemed to get beyond the individual and sub-unit level, including highly questionable "smartening up" drill and ceremonial parades. At the levels of brigade and battle group, lack of knowledge about how gunners operated and a misunderstanding of the benefits of artillery fire ensured the less than adequate integration of artillery with infantry and, more particularly, with tanks. As Colonel Stacey suggested in *Victory Campaign*, Canadians probably did not get as much out of their long training as they might have. In fact, one conscientious major who later rose to command a division, ended up in No. 1 Neurological Hospital ("No. 1 Nuts") because of psychological problems brought on by personal concerns that he was not receiving the training necessary to enable him to lead men in battle.

33

In many ways McNaughton was hoisted with his own petard during Exercise "Spartan" in March 1943 by his failure to treat the profession of arms as a discipline worthy of serious study. In contemplating passing one corps through another in the middle of the night, a difficult enough problem for brigades to accomplish, he demonstrated gross ignorance of large formation operations. He ordered 2 Corps at 2335 hours on 6 March to advance east; the next day at 1615 hours he issued a counter-order for a westward deployment that night; at 2130 hours on 10 March he issued orders for operations the following day; and at 2259 hours 11 March he issued orders for operations on 12 March. He appeared to have no idea that a corps required at least twenty-four hours warning in order to execute a major task. By way of rough comparison, US General Omar Bradley in ordering the redeployment of VII (US) Corps from the Cherbourg area in 1944 gave Lieutenant-General J. Lawton Collins five days for the turn-around of his corps: one day for rest, two for the move, another for reconnaissance, and a fifth on which to issue attack orders. Bradley called carrying out the redeployment a tall order for "Lightning Joe" Collins.

What McNaughton failed to appreciate was that an army commander should always be looking well beyond operations currently in progress and giving orders, not hours, but days in advance. To British General Alan Brooke, who visited "Spartan" on 7 March,

McNaughton simply "had no idea how to begin the job and was tying up his force into the most awful muddle." Sir James Grigg, British Secretary of State for War, reported that he "was appalled" at McNaughton's indecision as "he stood in front of his situation map hesitating as to what to do and what orders to issue." Even McNaughton's protégé and future successor, H.D.G. "Harry" Crerar, recalled "that it became patently obvious to all" during Exercise "Spartan" that McNaughton "was totally unsuitable for high operational command." According to Crerar, both Grigg and Ralston, then Canadian Defence Minister, personally observed McNaugton's "inability to sum up the situation and issue clear and precise orders." Not surprisingly, given such a track record, McNaughton's corps and division commanders were no better.

The performance of Major-General Basil Price, GOC 3rd Canadian Infantry Division, during Exercise "Beaver IV" in 1942 was so "lamentable," in Montgomery's words, that it confirmed he was a "complete amateur, one totally unable to train his Division." Before Price was eventually relieved, Montgomery indiscreetly wrote to his Canadian aide on course in Canada, "I hope to be sending Price back to you; he will be of great value in Canada where his knowledge of the milk industry will help on the national war effort." Similarly, Lieutenant-General Ernie Sansom, GOC 2 Corps, performed poorly during Exercise "Spartan" and even split his two armoured divisions, each made up of a tank and an infantry brigade trained to fight together, by grouping the two infantry brigades designed to support tanks under one armoured division and the two armoured brigades under the other. Although McNaughton quite correctly countermanded Sansom's doctrinally suspect regrouping, his action came too late to prevent it. In the ensuing traffic snarl-ups, tanks were left without infantry support and notionally destroyed in the pandemonium that reigned. This did not prevent Sansom, however, from preparing an adverse report upon Major-General C.R.S. Stein, GOC 5th Armoured Division, who was later summarily relieved in October of 1943 for "progressive anxiety neurosis." Sansom himself departed with McNaughton whose poorly timed attempt to visit Canadian troops in Sicily in July 1943, denied by General

Montgomery, sparked a minor nationalistic furor that perhaps as McNaughton intended overshadowed his lacklustre performance on "Spartan." It is significant to note in this whole sorry affair, moreover, that Crerar who agreed with the firing of Price, had as CGS originally appointed both him and Sansom to their positions as GOCs. Indeed, Crerar had also written to McNaughton in March 1941 that, "I think these are good appointments and should result in credit to the Commands as well as to their Commanders."

Between McNaughton's departure in December 1943 and Crerar's arrival as General Officer Commanding-in-Chief in March 1944, the First Canadian Army drifted essentially leaderless. The newly appointed acting commander, who was double-hatted as Chief of Staff, Canadian Military Headquarters, refused to get involved in operational detail. Canadian politicians, meanwhile, intervened freely in military affairs. Partly in response to calls to get Canadian soldiers into action for reasons of honour and postwar prestige, and partly to reduce the adverse effect of continued inaction upon troop morale, the Canadian government in 1943 arranged to have the 1st Canadian Infantry Division and 1st Canadian Army Tank Brigade replace British formations slated for the assault on Sicily. The original intent of Canadian participation in Italy was to have the soldiers from these groups return from the Mediterranean to disseminate battle experience throughout the First Canadian Army in preparation for the invasion of France. The post-Dieppe perception that casualties were likely to be lighter in Italy, however, prompted the King government to pressure British military authorities to accept a superfluous 1 Canadian Corps headquarters and an unwanted 5th Canadian Armoured Division for service in the infantry-consuming Italian theatre. The effect of splitting the First Canadian Army speeded the resignation of McNaughton and, in turn, relegated that formation to a "follow-up" as opposed to assault role in the invasion of Europe (which King had never warmed up to). The Canadian corps in Italy would not even fight as a corps, however, before the Canadian government commenced agitating in May 1944 for its early repatriation to the First Canadian Army.

As in the Great War, the field leadership of the Canadian army

took some interesting turns that could not help but affect troop performance. On the death of the commander of the 1st Canadian
Infantry Division in a plane crash, Lieutenant-General Guy Simonds
in April 1943 assumed his place and went on to become the most
battle-experienced senior field commander in Canada's overseas
army. During the course of the Italian campaign, Simonds managed
to impress Montgomery, who also owed his Eighth Army command
to the death of a predecessor in an aircraft accident. Simonds admired
Montgomery, but he was clearly not as impressed with McNaughton.
On learning that the latter proposed to visit Italy, Simonds blurted to
Montgomery, "For God's sake keep him away!" Simonds' relations
with Crerar also deteriorated over a "caravan incident" that saw a
busy Simonds eject a maintenance captain bent on obtaining the
exact measurements of his three-ton office van and sleeping quarters.
The captain worked for Crerar, who obviously wanted a van like
Simonds'. Crerar chose to make an issue of the affair by expressing
doubts to Montgomery about Simonds' suitability for high command.
When Montgomery backed Simonds, Crerar called in his chief psy-
chiatric adviser to examine some of the less than deferential correspondence Simonds had penned on the matter. The hapless adviser,
for his part, concluded that Simonds was sane but egocentric.

Arguably, Simonds was second only to Currie in the pantheon of
outstanding high-level field commanders produced by Canada. Like
Currie, he was highly respected for his competence despite his hot
temper and icy demeanour. Simonds never inspired troops in the
manner of his master, Montgomery, but by all accounts would have
gotten on well with the likes of the operationally brilliant yet severe
Field Marshal Erich von Manstein. Montgomery thought Simonds
easily the equal of any British corps commander. Recent evidence
also indicates that he considered Simonds and Sir Brian Horrocks to
be the two best corps commanders in 21st Army Group.
Montgomery's ideal army would have included their two corps,
which is exactly what he gave Crerar for the battle of the Rhineland.
Sir Miles Dempsey, GOC-in-Command of the Second British Army,
additionally considered Simonds first-rate and he worked well with
him. Such comments, even given their sources, are difficult for any

observer to ignore. That Simonds was only forty-one when he assumed command of 2 Canadian Corps in January 1944 is equally significant, for the fifteen-year difference that separated him from Crerar amounted to a half-generation age gap that pointed to a lost cohort of competent senior Canadian commanders who should have been trained during the inter-war years. In comparison, the youngest American corps commander was seven years older than Simonds, and of German corps commanders less than two per cent were under forty-five. Yet, there was no natural connection between youth and superior performance in the Canadian Army; the critical factor was knowledge of how to conduct military operations effectively in war. Unfortunately, the fateful 2 Canadian Corps effort to exploit the hard-won success of Operation "Atlantic," launched in support of Operation "Goodwood," a major British armoured attack on 18 July 1944, saw Canadian troops run away largely because they were placed in an untenable situation. Poor tank-infantry cooperation and the arrival of German reserves, especially 1st SS Panzer Division, which mustered in the neighbourhood of 100 tanks, merely sealed the Canadians' fate. The attack on Verrieres Ridge during Operation "Spring" on 25 July, which resulted in the virtual annihilation of the Black Watch, also sullied Simonds' reputation. But in the main, judging from their associated artillery fire plans, both of these actions were *divisional* as opposed to *corps* executed operations. To a large extent, Simonds was let down by his divisional commanders. Lieutenant-General Charles Foulkes, the GOC 2nd Division, not only mixed up his manoeuvre brigades, when they should have fought together as they had trained, but never went forward to get a grip on the battle. Yet, while most actions foundered on the shoals of inadequate tank-infantry and artillery coordination exacerbated by faulty divisional battle planning, the striking success of the Royal Hamilton Light Infantry (RHLI) in taking and holding Verrieres ultimately attested to what could have been done.

After two such bloody noses it is not surprising that Simonds chose to mount a major corps attack, if only to show his subordinates how to implement a proper fire plan. In Operation "Totalize," a daring night armoured attack launched on the anniversary of the

Canadian Corps Great War victory at Amiens, he actually cracked the German line wide open through the innovative use of improvised armoured personnel carriers (APCs) in a daring night operation launched behind a rolling barrage. Failure to realize that 1st SS had slipped away to Mortain regrettably caused Simonds to wait for a strategic bomber strike on where he wrongly presumed the Germans to be in depth. To make matters worse, the 1st Polish and 4th Armoured Divisions slated to exploit success, failed in the end to make any headway. The 4th Armoured, in particular, did not make sufficient use of its allotted artillery to blast its way forward. The commander of the spearhead 4th Armoured Brigade, who possessed adequate means to coordinate intimate artillery fire for units, also abrogated responsibility for his tank-infantry battle groups, neither of which possessed as good a means of coordinating artillery fire. He was later found kilometres away asleep in his tank, and possibly drunk, by the division commander, who himself could have done much more to "stage manage" the operation. In one of the few 12th SS actions in Normandy where panzers played a greater role than antitank guns "Totalize" was subsequently halted. We are thus left to ponder that had 2 Canadian Corps in the early morning of 8 August raced immediately toward Falaise, the entire Seventh German Army would have been caught in full stride counter-attacking west toward Mortain, unable to avoid the utter calamity of a stormtight encirclement.

Simonds was not, of course, the first senior Canadian formation commander to fight in France. That honour belonged to Major-General R.F.L. Keller, GOC of the 3rd Canadian Infantry Division, which landed on D-Day as part of 1 (British) Corps in the Second British Army. As a physical training instructor at the Royal Military College, Keller had earned the nickname of "Captain Blood" and from there he went on to acquire a reputation as a physically hard, tough-talking officer. We know that Crerar in 1942 entertained great confidence in Keller's "two-fisted and competent" field leadership and considered him corps commander material as late as May 1944. Others, subordinates as well as superiors, were not so sure. As reported by the GOC 1 British Corps, Sir John Crocker, Keller was not really fit temperamentally or perhaps even physically for high command (he had the appearance of a man who had lived pretty well, observed Crocker). With the exception of one well-led brigade, Crocker continued, Keller's division had "lapsed into a very nervy state after the excitement of the initial phase had passed." The jumpiness of the division, recorded Crocker, "was a reflection of the state of its Commander [who] was obviously not standing up well to the strain and showed signs of fatigue and nervousness (one might almost say fright) which were patent for all to see." In both Operation "Windsor," the attack upon Carpiquet, and Operation "Charnwood," the fight for Caen during the Normandy campaign, Keller failed to provide the necessary direction and guidance expected of a division commander. A subsequent fight for Tilly-la-Campagne resulted in the dismissal of one of Keller's brigadiers and two of his unit commanders, when perhaps it should have been Keller who went. Fortunately, Brigadier Todd, the commander of the divisional artillery (CRA), managed to take up the leadership slack reasonably successfully until Keller's wounding and eventual replacement by Major-General D.C. Spry.

As for Crerar himself, he too was described as "austere and frigid," respected but unloved, with "some element of terror in the manner of his command." Colonel Stacey personally witnessed Crerar tear a strip off his Auster aircraft pilot to the great embarrassment of a large number of bystanders within earshot. He also got into a silly fight with Sir John Crocker, his one and only corps commander when the First Canadian Army initially became operational in Normandy. In some exasperation, General Montgomery wrote to General Brooke: "Harry Crerar has started off his career as an Army Comd by thoroughly upsetting everyone; he had a row with Crocker the first day, and asked me to remove Crocker. I have spent two days trying to restore peace.... As always, there are faults on both sides. But the basic cause was Harry; I fear he thinks he is a great soldier, and he was determined to show it the very moment he took over command at 1200 hours 23 July. He made his first mistake at 1205 hr.; his second after lunch." To several Canadian historians, including Stacey, this dispute was simply a matter of insubordination and had Crerar been a British officer, he would not have had to go to Montgomery to get

Crocker to obey. Actual map examination of the tactical issues involved, however, indicates that Crocker had reasonable grounds on which to remonstrate as a corps commander. Crerar's solution to clearing the Orne Canal of direct enemy observation and fire, which he should properly have let the corps commander decide anyway, could well have resulted in major losses without necessarily accomplishing the task. In any case, much as the more experienced Crocker suggested, two different operations other than the one envisaged by Crerar were eventually planned.

Although the ambitious Crerar retained command of the First Canadian Army to the end of the war, he never really shone as a major field commander. During Operation "Totalize," he was the only one who could possibly have turned off the bombers in spite of Simonds's protestations, but for evident reasons related to comparative military knowledge, this he dared not do. Dempsey, in contrast, rarely hesitated to overrule his corps commanders when he thought the operational situation dictated. After all, one might ask, what was the use of having an army commander? Crerar's ponderous clearance of the Channel ports and coastal belt on the way to the Scheldt has additionally been much criticized. Indeed, in the view of two Canadian military historians, operations were conducted so poorly that had Crerar been a British rather than Canadian general, he would most likely have been sacked as an army commander. Simonds had also supposedly protested against Crerar giving priority to the physical clearance of the Channel ports, and suggested instead that 2 Canadian Corps conduct a relentless pursuit along the coast to Breskens, Holland, masking German coastal defences encountered, and then turn east to cut the Fifteenth German Army off in the Scheldt estuary north of Antwerp. To ease transport difficulties, Simonds further proposed provisioning his corps with ammunition, fuel, and bridging by beaching preloaded landing craft at designated intervals during the advance.

Simonds finest hour came during the Battle of the Scheldt when he assumed temporary command of the First Canadian Army on the medical evacuation of Crerar for persistent dysentery. In a series of bitterly contested actions conducted under the most appalling conditions, including a brilliant waterborne attack by the 3rd Canadian Infantry Division under Major-General Dan Spry[1] the First Canadian Army succeeded in opening the great Belgian port of Antwerp. The struggle for the Breskens Pocket in the Scheldt estuary cost almost 2000 Canadian casualties, over 300 of them dead. Neuro-psychiatric or battle exhaustion cases accounted for 16 per cent of the infantry total, and, of these, about 90 per cent had three months or more of combat experience. By 17 October, as a result of heavy fighting at Woensdrecht, Holland, the average rifle company strength within the RHLI (Royal Hamilton Light Infantry) plummeted to forty-five and two out of four companies had only one officer. Such losses approached those sustained by the Canadian Army in Normandy where the 3rd Canadian Infantry Division incurred more casualties than any other division in the 21st Army Group. The 2nd Canadian Infantry division, though it did not arrive in France until 7 July, suffered the next highest. By mid-August infantry losses comprised a shocking 76 per cent of total army casualties. Field personnel returns for 26 August showed six of the nine infantry battalions of the 2nd Canadian Infantry Division to be short more than 200 men. By early September 1944 average battalion strength within the division had fallen to 525. In the Battle of the Scheldt platoon strengths that were established at around thirty-four dropped to sixteen men. A few companies mustered twenty-five.

Since the government of Mackenzie King had steadfastly refused to send conscripted personnel overseas, the Canadian Army in some desperation resorted to a policy of remustering volunteers to redress infantry shortages. As one veteran commented, the army was "forced to practice self-cannibalism in order to keep the fighting units in existence." Infantry reinforcements arriving in the Scheldt thus included surplus engineers, gunners, tankers, ordnance, service corps, and even air force ground crew, in addition to drivers, clerks, cooks, and other tradesmen. As became painfully apparent, however, many of the reinforcements had been transferred to the infantry with only the most rudimentary training. Some did not even know how to strip and assemble a Bren gun, while others had not seen a PIAT[2] or ever thrown a grenade. Within the four rifle companies of

41

the Black Watch on 19 October, 174 out of 379 all ranks had not received more than a month's training. In another battalion a Canadian sergeant complained, "we got five or six butchers, but they were the wrong kind." While some fighting units instituted crash training programs to prevent such soldiers from becoming a danger to themselves and their comrades, others had no alternative but to assimilate them in battle. Meanwhile, roughly 70,000 home defence conscripts, many with extensive infantry training, lived relatively comfortably in barracks in Canada.

The Canadian public first learned about the overseas infantry reinforcement problem when a prominent hockey promoter and wounded returnee, Major Conn Smythe, MC, broke the story to the press in Toronto on 18 September. In a signed statement Smythe declared that enough evidence existed to suggest that poorly trained reinforcements were causing "large numbers of unnecessary casualties" in combat. His conclusion that well-trained conscripts should now be sent to the battlefronts not only received broad public support, but also prompted Defence Minister Ralston to depart shortly on a fact-finding mission to overseas theatres. After a tour of the Italian front and a four-day visit to the First Canadian Army, Ralston returned to Canada on 18 October convinced of the need for overseas conscription. At a cabinet meeting on 24 October he explained that as the early collapse of Germany could no longer be expected, he saw no alternative but to recommend sending home defence conscripts overseas. In the ensuing crisis King manoeuvred to replace Ralston with McNaughton who still believed that he could breathe new life into the volunteer system and thereby avert the need for overseas conscription. King then called another cabinet meeting for 1 November and, with McNaughton waiting in the wings, bizarrely accepted a letter of resignation that Ralston had previously tendered in July 1942 during the first conscription crisis (which released King from his original promise not to conscript for overseas service). By 22 November, however, it was obvious that the volunteer system could not provide the necessary overseas reinforcements. The next day, faced with the prospect of either ordering conscripts overseas or seeing his government break up, Mackenzie King rose in the House to read an order-in-council authorizing the dispatch of home defence conscripts to Europe.

That the King government faced its greatest crisis of the war at the height of the Battle of the Scheldt is not without irony. The Canadian Army had always been the Cinderella service and the impression is hard to escape that the Prime Minister would personally have preferred Canada's volunteer soldiers overseas to have fought and died with as little fuss and bother as possible. From the beginning the King government had attempted to fight a war of "limited liability," concentrating on industrial production and air, naval, and garrison[3] forces so as to obviate any requirement for fielding an overseas army or conscription. King himself apparently believed, wrongly as it turned out, that an air commitment would involve fewer casualties. In the end, however, government defence policy backfired and Canada entered the war in such haphazard fashion that it inadvertently condemned the RCAF to colonial status, created two classes of soldier within the army, and denied itself any strategic voice. Incredibly, King's senior military advisers did not begin to attend Cabinet War Committee meetings on a regular basis until the war was nearly three years old, and then only the first and third meetings of each month. The result was that the War Committee collectively tended either to neglect or to over-manage military affairs, usually subjecting them to the most excruciating examination often from a mainly domestic perspective. The September 1941 transfer of the major part of the RCN to *neutral* American naval command, under whose operational control it remained until the Atlantic Convoy Conference of March 1943, passed practically unnoticed. The Japanese shelling of Estevan Point on Vancouver Island in June 1942, on the other hand, saw the CGS heading west to take personal charge of coastal defences (to the point of telephoning an order for six Bofors guns from Prince Rupert), when he most properly belonged at the seat of government to proffer military advice.

In contrast with Winston Churchill and Franklin Roosevelt, both of whom leaned heavily upon their military advisers, King never appreciated the presence of the Canadian COS. Given his continuous obsession with conscription and domestic affairs, it is probably not surprising that Canadian fighting soldiers overseas began to consider themselves a forgotten army. Within their ranks there also arose the phenomenon of "de-Canadianization," the outright antagonism of many troops toward their government's policy and their countrymen at home. It is sobering to reflect, as well, that thrice-wounded soldiers were often sent back into the line. Until the fall of 1944, moreover, no system of rotational leave existed for Canadian soldiers overseas, and then only those with five years continuous service overseas qualified. When many eventually returned home on demobilization, their children were six years older. Yet, it is also true that throughout its long ordeal this castaway army never cracked nor went to pieces in the field. By the end of the war, having paid a steep price in blood for the peacetime neglect of military professionalism, it was probably the best little army in the world. Certainly, in the performance of the Canadian Army overseas the government of Canada got much more than it deserved. However, one doubts whether the Canadian people, whose sons bravely bore the brunt of the firestorm, really got as much as they should have from the peacetime endeavours of the professional military establishment they retained during the interwar years. Surely there are still lessons to be learned here by politicians and soldiers alike.

NOTES

1. Although Simonds would later fire Spry after the battle of the Hochwald Gap, there is reason to believe that this youthful general by that time rated among the top three divisional commanders produced by Canada. The others were Major-Generals H. W. Foster and B.M. Hoffmeister, formerly of "No. 1 Nuts," who may have been the finest. Of course, it is difficult to make such assessments without rigorously examining their respective performances in detail. Hoffmeister unquestionably performed well in Italy, and even brilliantly in the breaking of the Gothic Line, which may well rank as Canada's finest victory in the Second World War. It is nonetheless significant as well as indicative of the Canadian Army training system to note that Hoffmeister, who had yearned for instruction that wasn't there in Britain, did

not think he could have commanded a battalion effectively without the benefit of Staff College training.

2. Projector Infantry, Antitank.

3. Canadian forces garrisoned Bermuda (1940, 1942-6), Jamaica (1940-6), Nassau (1942-6), British Guiana (1942-5), and Hong Kong (1941-2).

Chapter Four

The Flowering of Military Professionalism

From most indications, the Canadian Army after the Second World War sought to institutionalize military professionalism to save it from the erosion that afflicted the CEF. Canadian General Staff post-war plans called for retaining a regular force of 55,788 all ranks and recruiting a reserve militia of 177,396 personnel through a universal system of military training. In October 1945, however, the Minister of National Defence, Douglas Abbott, announced that the permanent active force would tentatively comprise between 20,000 and 25,000 all ranks. In addition to fielding a "ready" brigade group of regulars, the permanent force was to train and administer a part-time volunteer reserve force organized in six divisions with ancillary troops for an army of two corps. In fact, Abbott's main task was to oversee demobilization. By 1947[1] the regular army had been reduced to 14,185 troops in two small armoured regiments, three under-strength infantry battalions, and about eight batteries of artillery. The next year witnessed an increase in regular force strength to just over 20,000, while the authorized ceiling of the neglected reserve force militia hovered around 43,000. In 1952 only 3908 officers and 9865 militiamen attended summer training camp. Although the militia paraded two-thirds of its officer establishment in 1953, it fielded only a fifth of its establishment in men, most of whom were senior NCOs.

Abbott, who was also Minister of National Defence for Naval Services as well as army minister, went on to become Minister of Finance in 1946. The departure of Colin Gibson as Minister of National Defence for Air at the same time enabled the Department of National Defence to be consolidated under one minister, Brooke Claxton. Claxton served as defence minister for eight years and it was said of him that not since Sir Frederick William Borden had a peacetime defence minister accomplished so much. Unquestionably, Claxton shaped defence policy and, although he was a workaholic with acutely sensitive political antennae, he retained the unshakeable conviction that his primary duty was to maintain a viable peacetime defence establishment. In addition to integrating the three service headquarters and creating a single civil service within the defence department, he streamlined defence administration, introduced an updated National Defence Act, reopened the Royal Military College on a tri-service basis, and established a Defence Research Board and National Defence College. During his tenure, he also promoted major land force exercises in the north, participated in the formation of the North Atlantic Treaty Organization, weathered the Korean War, and set in train the construction of radar warning systems in the Canadian north. In 1953 by his forthright honesty in Parliament, he further managed to defuse a politically explosive scandal related to the Currie Commission's investigation of misappropriated army construction funds at Camp Petawawa.

Claxton's last significant initiative was to create the post of Chairman of the Chiefs of Staff Committee replete with a permanent secretariat. Charged with responsibility for coordinating the operations and training of the Canadian forces, the Chairman was the minister's principal military adviser and the senior Canadian military officer. When Claxton promoted Lieutenant-General Charles Foulkes to this position in 1951, however, he also selected Lieutenant-General Guy Simonds to head the Canadian Army as Chief of the General Staff. From the perspective of fostering military professionalism within an expanding army, the latter was undoubtedly the more astute move. Foulkes had originally been chosen by Abbott for his diplomatic and political acumen that made him more amenable to compromise. In fact, Foulkes was the consummate bureaucrat who was always prepared to compromise on issues to maintain a cordial relationship with his political masters. By the end of his fifteen-year tour in Ottawa, he reputedly changed direction with the wind. The outspoken Simonds, in contrast, was no diplomat, but he was the better field soldier and possessed of a more imaginative and creative intellect. He was also arguably the most effective CGS the army ever had and his legacy of professionalism continued under his successors. One of his former subordinates, Howard Graham, served as CGS from 1955 to 1958; his former corps chief signals officer, Findlay Clark, from 1958 to 1961; and his former corps chief engineer, Geoffrey Walsh, from 1961 to 1964.

During the rapid peacetime expansion of the forces from 1951, which year saw a hefty 6.6 per cent of GNP devoted to defence, Simonds instituted a high standard of military training and education within the Canadian Army. Having been the first commandant of what became the Canadian Army Staff College (CASC), chief instructor of the British Imperial Defence College, and first Canadian commandant of the National Defence College in Kingston, Ontario, he was ideally qualified to shore up the institutional foundations of Canadian military professionalism. The challenge was to create a peacetime officer corps imbued with character and brains. Although he was himself an RMC graduate who opposed the closing of the college in 1942, Simonds harboured deep reservations about the ability of that institution to produce the right sort of subaltern (junior officer), especially as it looked as though tenured civilian academics might eventually become too influential.[2] As Simonds correctly perceived, perhaps more so than most officers, staff colleges were the true nurseries of the General Staff and had supplanted cadet academies in the advancement of military education. He accordingly assigned the highest priority to the development of the CASC, doubling the size of the student body and improving its facilities and faculty. His appointment of Brigadier George Kitching to the commandantcy in 1951 was also a stroke of genius as Kitching was probably the brightest staff officer in the army and had learned from his experiences in war. His successor after 1954, Brigadier Pat Bogert, was also described as one of the most intellectual officers in the army.

However highly recommended by a superior, an officer before attending staff college had to pass a battery of stiff formal qualification examinations at lieutenant and captain rank. Meticulously administered and comprising separate comprehensive written and *practical* portions designed primarily to test an officer's fitness for war, these examinations accorded with the old German army principle of training two rank levels up. An officer was expected to be ready to attempt "lieutenant to captain" examinations by not later than his third year of commissioned service and "captain to major" examinations by not later than three years after promotion to captain. Preparation for and application to write each qualification examination was the responsibility of the individual officer. In each case, the written theoretical portion comprised five examinations covering such subject areas as military law, military history, current affairs, staff duties and military writing, organization and administration in peace and war, and tactical operations. To ensure complete fairness by centralized marking boards, candidates were forbidden to identify themselves by either name or regiment. The practical field portion, which could only be attempted upon successful completion of all written examinations, was further sub-divided into a "common to all corps" section (mandatory for all officers) and a "special to corps" section pertaining to an individual's specialty. To pass these promotion examinations with a "distinguished" mark meant that one's advancement could be accelerated. Failure to pass, on the other hand, meant that one would not be promoted. To fail three times meant that one would never be promoted again. Until 1959 prospective officers were also required to pass a staff college entrance examination; thereafter, they were selected on the basis of high grades attained on "captain to major" examinations.

Apart from reinforcing the institutional foundations of a military professionalism that possessed substantial intellectual dimensions, Simonds also convinced Claxton of the importance of producing effective mobilization plans. The Korean War mobilization had not gone well largely because of the absence of such plans and a public outcry for action that spurred too hasty a ministerial intervention and a resultant lowering of enlistment standards. Having opted to retain the regular army for defence of Canada operations, the government decided to recruit from volunteers in the militia and civilian sectors a completely new Canadian Army Special Force to fight in

49

Korea. This force, organized as the 25th Canadian Infantry Brigade, was commanded by a distinguished war veteran and militiaman, Brigadier J.M. Rockingham. Despite the initial confusion and difficulty in raising the brigade, the Canadian force acquitted itself well.

In recruiting the 27th Brigade which was slated for service with NATO in Europe, the Canadian Army again turned to the militia. Simonds, always a believer in the power of the militia to find men if they could continue to be affiliated with their units, proposed creating six regionally diverse composite battalions from fifteen militia regiments each providing two companies. In the event of war, he reasoned, these companies could be expanded into a battalion, thereby providing the basis for a field force of two divisions. By naming a Major-General Reserves to sit alongside his Vice-Chief of the General Staff, Adjutant-General, and Quartermaster-General on Army Council, Simonds provided a finishing touch to the integration of reserve and regular forces. Eventually, however, owing to administrative difficulties arising from different terms of service among Korean, NATO, and home brigade soldiers, all active duty units were made regular. In 1952 for the first time in Canadian history regulars outnumbered reservists. The decade after the Korean War also witnessed the development of a Canadian army doctrine that represented the apogee of a peacetime military professionalism based on the lessons of the two world wars and Korea.

NOTES

1. In December 1944 the Cabinet War Committee had approved an army occupation force of about 25,000, but only 18,000 men of the 3rd Canadian Infantry Division actually served under 30 British Corps in that capacity. In the spring of 1946 this Canadian Army Occupation Force, afflicted by sit-down strikes by soldiers anxious to get home, was withdrawn as part of a general demobilization.

2. True to his convictions, Simonds refused to accept an honorary degree from RMC.

Chapter Five

Long Journey into Night

With the benefit of hindsight it is now possible to see that the post-Korean War Canadian army suffered two major blows from which it never fully recovered. The first was absorbed mainly by the militia and reflected both the emphasis placed on the aerial deterrence of war in the era of "massive retaliation" proclaimed by US Secretary of State John Foster Dulles in January 1954 and the growing fear of nuclear attack. One Canadian manifestation of this trend was that the strength of the RCAF surpassed that of the regular army in 1955. Another manifestation was the move toward using the militia in national survival operations.

In June 1954 Claxton commissioned a senior officer board headed by Major-General Howard Kennedy to examine and report upon the organization, training, and administration of the Canadian Army reserve force. A key recommendation of the board was that the role of the reorganized reserve force, henceforth to be called the Canadian Army (Militia), was to provide a partially trained and equipped force as the nucleus of units to be mobilized and brought up to strength in case of emergency. By the end of 1954, however, Major-General F.F. Worthington, Civil Defence Coordinator under the Department of National Health and Welfare, which in 1951 had taken responsibility for most civil defence functions from the Department of National Defence, was calling for the militia to be used in national survival search and rescue operations in the event of nuclear attack. Simonds, worried that such a specialized role would turn the militia into a safe haven for draft dodgers, responded that armed forces maintained and trained primarily for warfighting could always offer assistance to the civil defence authority. Meanwhile, although the reinvigorated militia experienced a marked revival in summer camp training between 1955 and 1957, all such training ceased thereafter, ostensibly to free up money for the purchase of equipment for the regular force.

In 1958 during the so-called "golden age" of deterrent thought that accentuated "forces in being," defence minister George Pearkes confirmed that the primary role of the militia would henceforth be to restore order and conduct search and rescue re-entry operations into target areas in the event of a nuclear attack on Canada. The following year responsibility for most civil defence functions, apart from humanitarian tasks, reverted to the Department of National Defence. The result was that the army at large inherited the civil defence role. By the spring of 1960 the regular army fielded 22 mobile support columns, backed up by 44 mixed columns of regular and militia troops, for what were now officially termed national survival operations. But whereas the regular force continued to conduct conventional field training, survival training remained the first priority of the militia. Indeed, some militia artillery units complained that they were unable to produce a four-gun troop capable of firing. Only at the end of 1963, when the government slashed militia strength from 51,000 to 32,000, did the emphasis accorded national survival begin to subside. In 1964 a ministerial committee headed by Brigadier E.R. Suttie listed militia roles in descending order of priority as: bringing the regular field force up to war establishment in time of need; providing trained reinforcements for overseas formations; maintaining internal security; and assisting the regular force in national survival operations.

The second, more devastating blow landed on the army came gloved in the government White Paper of 1964, which placed top priority on peacekeeping and ushered in a generation of professional decline within the army. In August of that year, in an attempt to realize its new goals for the armed forces, the government replaced the three headquarters of the separate services with a single integrated Canadian Forces Headquarters (CFHQ) with four staff branches. Within CFHQ the former Chief of the Air Staff assumed the post of CDS with the former CGS as his Vice-Chief. At the same time, the separate administrative, supply, and personnel organizations of the RCN, Canadian Army, and RCAF were integrated to form single service branches. Although many informed observers applauded these actions as long-overdue cost savings, the ramifications of the actual steps taken to effect service integration were not entirely fathomed. The adoption of the air force's functional command[1] and base system in 1965-66 left much to be desired since operational roles, except for anti-submarine coastal defence, could not be integrated. The mobile, global strike force envisioned by the ambitious new Defence Minister Paul Hellyer for peacekeeping and brushfire wars was largely stillborn for lack of sealift and capital equipment expenditures. What was also not well understood was that the General Staff was uniquely integral to the field operations of the army and only in the broadest sense was it similar to the more centralized agencies of the RCN and RCAF. In fact, the army lost a tried and proven General Staff system that had provided strong professional linkages between the high command in Ottawa and army elements operating in the field. What previously constituted the army "capital" or central planning staff, as opposed to the army operational staff with field troop formations, could also only be faintly discerned in CFHQ. To Colonel C.P. Stacey, "the rats had got at it."

Worse yet for the army, the system for training General Staff officers underwent substantial modification. A new tri-service Canadian Forces Staff College (CFSC) based on the former Air Force Staff College and emphasizing executive development superseded the CASC, which under the less imposing name of Canadian Land Forces Junior Staff College conducted truncated versions of the old army staff course. Although the continued professional worth of the latter college came to be universally appreciated in time, despite continual attempts to close it altogether, the high standards formerly set by the Canadian General Staff were never equalled. Unquestionably, the termination of army qualification examinations in 1966, *more than any one other factor*, accounted for this professional regression. It was the equivalent of the legal profession dropping bar examinations. As entrance to staff college no longer depended on achieving high marks in "captain to major" examinations, educational standards had no place to go but down. In fact, the discontinuance of formal competitive theoretical and practical examinations removed the most reliably objective means of determining an officer's fitness for promotion. In an army where anti-intellectual currents always lurked beneath the surface, the examinations had also promised to ensure the steady progressive intellectual development of the peacetime officer corps. With the end of examinations, officers merely had to please their bosses and jump through prescribed career hoops. Many who could never possibly have gained entry to staff college in Simonds' time were given the opportunity to attend. As staff college graduation was also a prerequisite for promotion, numbers of these less qualified officers went on to attain senior rank, some rising as high as lieutenant-general.

The highly centralized unification of the forces in February 1968 delivered further rude shocks to an already reeling army. The striking of Simonds' guards, rifle, and highland battalions from the regular force establishment, because they looked too British, contributed to a growing troop disenchantment that eroded morale and esprit de corps. The issue of new bottle-green uniforms, which some soldiers thought made them look like bus drivers, did little to allay fears that most things of importance were being reduced to some lowest common denominator (one CDS, as if to prove the point, donned a specially designed green uniform that made him look like a South American dictator). But whereas Hellyer was able to get rid of the RCN and the RCAF with relative ease, he could not eliminate the British regimental system that had taken root so firmly in the soil of Canada. In many respects this was understandable, since the redeeming features of the regimental system—the promotion of group cohesion and fighting spirit—had usually more than compensated for its intrinsic shortcomings of fostering of an extreme parochialism and narrowness of view that stultified professional thought. Fortunately for Canada's army, it was also the much maligned and neglected militia with units spread throughout *voting* communities from coast to coast that saved the regimental system.

In a regrettable turn, however, regular force regiments no longer constrained by a properly functioning General Staff gradually arrogated power unto themselves. Generals who rose from their ranks often aided and abetted their regiments in this trend. Soon many of them were acting like petty Mafioso, with regimental godfathers cutting deals to ensure the advancement of favoured sons to selected positions whether they were qualified or not. The iniquity of this development was that the brightest person was not always chosen for the job. Senior appointments were often made largely because it was this or that regiment's turn and the anointed nominee needed to have his ticket punched in a certain area of employment. A misplaced fixation on producing as many general officers as possible to advance regimental interests furthermore compelled the too rapid progression of many promising young officers. At the same time, in their headlong rush to the top they often blocked older, better qualified officers from assuming the command and staff appointments the young officers temporarily filled along the way. One adverse effect of rocketing officers through the ranks was that it encouraged a dilettantish rather than professional approach to soldiering. Another was that it encouraged a self-centred careerism nurtured by a subjective performance evaluation system that, despite denials, really boiled down to Lord Palmerston's snort: "Merit? The opinion one man entertains of another." Meanwhile, those kingmakers and dons who did little else than make or break regimental careers were usually so immersed in their personnel tinkering that they had little of value to offer the Canadian profession of arms.

All these developments took place in a climate of constant change and reorganization, often only for the sake of change. Trendy management techniques advocated by US Secretary of Defense Robert MacNamara and his whiz kids during the Vietnam War were embraced largely without qualification by Canadian military imitators in headquarters and training establishments. In 1967 Hellyer departed the defence portfolio leaving the havoc created by unification in his train. His successor, Léo Cadieux, seized upon the opportunity to launch yet another major initiative, namely, the promotion of bilingualism and the establishment of French language units in Quebec. In 1969 Prime Minister Pierre Trudeau, apparently without consulting DND, reversed the priorities assigned to the forces by Hellyer's 1964 White Paper. The protection of sovereignty now came first instead of last, with peacekeeping last rather than first. North American defence came second, ahead of Canada's NATO commitment due to be reduced and consolidated by 1970. To complicate matters further, the 1971 White Paper, which seldom used the word war, called for a Management Review Group (MRG) to examine streamlining coordination between military and DND civilian staffs in Ottawa. Out of this came a further headquarters reorganization, which saw the integrated CFHQ combined with the Deputy Minister's (DM) staff to form one restructured NDHQ with civilian and military assistant deputy ministers heading staff branches. Although the MRG had proposed that the DM outrank the CDS, the new structure stressed their equality in a duumvirate, with the former responsible for management and the latter for operations. Unfortunately, some DMs, though totally unqualified, were not above dabbling in nuclear deterrent theory and even operational matters.

Predictably, the new CFHQ before its demise in 1972 appeared to be incapable of managing itself and had little or no credibility in the field. A major reason for this deficiency, apart from the termination of General Staff linkages in the army case, was that for reasons of economy the headquarters had almost been refashioned into a fourth service mainly concerned with procurement. Little wonder, then, that such an astute observer as Professor Colin Gray concluded that the defence policy of Canada was simply equipment. The NDHQ designed by the MRG became even more immersed as a whole in financial management, personnel administration, and capital procurement, to the point where bureaucrats prevailed over operators. Regular meetings of Defence Management Committee, co-chaired by the DM and CDS, and the Program Control Board, chaired by the Vice-Chief, through their nomenclature as much as through their agenda confirmed a principal NDHQ focus on the management and control of capital expenditures. The general consensus, too, was that the DM, who took precedence over the CDS in address lists for internal correspondence, began to call the shots operationally as well as managerially. In recognition of this development, some among the more ambitious military bureaucrats even took to saluting him, though he did not warrant a salute, while others were overheard to say that they would rather follow the DM than the CDS. This was the same sort of obsequious behaviour commented on by the German essayist Heinrich Heine in 1841:

> Some years ago, just as I was making my way to Herr von Rothschild, a servant in livery carried his chamber pot across the corridor, and a stock market speculator, who was passing at the same time, removed his hat respectfully before the mighty utensil.... I noted the name of this devout person, and I am convinced that, in the course of time, he will become a millionaire.[2]

Meanwhile, although Prime Minister Trudeau had not succeeded in repatriating all Canadian forces from Europe in 1970, the authorized ceiling of the Canadian land contribution there dropped from over 6000 soldiers in a heavy brigade group with British Army of the Rhine to 2800 soldiers in a battle group sitting on two Canadian airfields in southern Germany. With only six rifle companies, three artillery batteries, two tank squadrons, and an oversized headquarters, the Canadian force was a barely viable fighting formation that often taught bad tactical lessons on exercise. The assignment of this force to a minor reserve role in support of either VII US Corps or II German Corps also left its commanders and staff to be courted by Americans and Germans as much as put to the test. In any case, whereas Canadian performance had previously been measured against the reasonably high professional standards of British Army of the Rhine, it could now only be compared against a US Army in decline and a culturally different German conscript force using another operational language. Once one of the most effective formations on the NATO central front, the Canadian army in Europe had, in fact, been reduced to one of the least effective[3]. Worse yet, noted a Canadian brigade commander in 1984, shortcomings in equipment, training and logistical support, and operational prowess had been concealed in a maze of unrealistic peacetime plans and perceptions. The addition of another tank squadron, an artillery battery, and two rifle companies in 1987 seemed almost too little too late. To one Canadian officer who observed that the formation could still only muster fifty-nine tanks, a distinguished German veteran commented,

"Well, you can't have a general then!"

But the truth was, the Canadian Forces suffered from an excess of generals. From 84 in 104,427 all ranks of the three services in 1953, they had multiplied to 107 in 79,383 by 1974 and to 112 in 67,000 by 1994. The decision to equate military ranks to civil service grades for purposes of pay and allowances had not ensured a commensurate rise in quality. The shameful practice of providing general officers with a better life insurance policy than those they were supposed to command in war additionally undermined their status as leaders obligated to share relatively equal hardships. Army officers who had voluntarily placed themselves on parachute courses, but then refused to jump at the head of their men, had also been allowed to enter the bloated ranks of generals. One, a lieutenant-colonel, even rose to brigadier-general within the halls of NDHQ where an airman had earlier set a bureaucratic promotion pattern by ascending from major to major-general. Although usually not so blatantly obvious, the transformation of generals into bureaucrats under the guise of executive development was an insidious feature of the senior officer promotion pattern within Canada's unified forces. At the same time, the advent of bilingualism opened the floodgates to further military mediocrity as it became more important to know French than to know one's job. In a particularly glaring case, the Combat Training Centre was handed to a francophone who was not only inarticulate in both official languages, but militarily incompetent as well. The effect on the field training system was disastrous and—as the little man busied himself with kingmaking—lasting.

As Canadian historian Professor J.L. Granatstein correctly perceived, the Trudeau years from 1968 to 1984 were a "long, dark night of the spirit" in which the Canadian Forces lost their way. But clearly, budget cuts and a combination of political neglect and interference were not entirely to blame for the forces' resultant predicament. The new Canadian land forces continued to bear the scars of an inordinate number of self-inflicted wounds. They alone reduced themselves to a collection of competing regiments. Notwithstanding the lessons of the Second World War, too many commanders preferred to act as exercise directors rather than test themselves. The awful practice of double-hatting also continued unabated, even to the extent of making the army staff college commandant the deputy commander of the 1st Division (which was also commanded by a non-combat arms general!). Meanwhile, the writing of doctrine came more and more to be delegated to junior officers, when it was no secret that generals of the stature of Montgomery had written their own manuals. During the same period, officer and NCO professional training in units generally degenerated in quality and frequency, while formal test exercises for commanding officers and formation commanders all but disappeared. Live firing training likewise ceased to be attempted as often as it once was, largely because careerist officers preferred not to risk blotting their copy book. As many of these careerist officers later reached higher rank, such professional knowledge was lost to new subordinates with often disastrous results in the conduct of live firing training. The constant reorganization of the training system did little to help arrest this professional slide. Between 1970 and 1983 the Combat Training Centre in Gagetown, New Brunswick, underwent seven major reconfigurations, or roughly one reorganization every two years. The conclusions of a 1972 study of professionalism hardly mentioned fighting skills as a first priority and said not a word about how to develop them. Subsequent efforts to turn the tide back with new uniforms and partly cosmetic troop concentrations that accentuated form over substance fell short of the mark as too much expertise had already been lost.

Adding to the turbulence of the period was the increased involvement of the Canadian army in peacekeeping. To a considerable extent the army became a tool of the External Affairs department, even though defence policy has never been synonymous with the diplomacy that armed force is *ultimately* expected to back up. By 1994 Canada, the self-proclaimed inventor of the genre, had sent more peacekeepers abroad—some 100,000 over four decades—than any other nation. Having participated in every United Nations peacekeeping mission since 1948, it almost became a matter of high principle to continue to do so. Yet, altruism was not what eventually motivated the Canadian army to actively embrace peacekeeping. Indeed, the old professional army had always been cool to the concept, especially after the debacle of the 1960-4 Congo operation that exposed the limitations of peacekeepers in resolving a seemingly endless tribal civil war. What gradually changed this army outlook was the subsequently ceaseless involvement of Canadian soldiers in peacekeeping to the point where many of them who should have known better began to accept it as a primary role. No national interest underpinned the commitment of Canadian troops to Somalia or Bosnia. Old-fashioned inter-service rivalry nonetheless prompted the army to seize upon the peacekeeping role as a means of acquiring a greater share of the defence budget. The army, proclaimed retired Major-General Lewis MacKenzie, "will be the Canadian Force's primary instrument of Canada's foreign policy into the next century and, therefore, deserves a higher priority for personnel and equipment within the Canadian military's budget than it has now." UN Secretary General Boutros-Ghali's assertion that peacekeeping had become a growth industry lent further encouragement and timely impetus to the establishment of a Canadian Peacekeeping Training Centre with defence dollars. But, here again, altruism and national defence concerns took a back seat to vested interests, for there were reputations and salaries to be made in the venture.

Part of the attraction of peacekeeping, aside from money (American units were the only troops not to take UN pay in addition to national pay), was that it afforded the soldier an opportunity for travel, adventure, and medals. But there was also a downside to peacekeeping from a professional military perspective. Although junior NCOs in charge of isolated outposts often faced greater leadership challenges than, say, commanding a tent group in the Arctic, the routine of peacekeeping could be an exceedingly boring and lonely business for extended periods of time. Maintaining real proficiency in battledrills, battlecraft, and other tactical training was practically out of the question for most units, and the non-coercive impartial nature of peacekeeping duty more often than not provided little scope for officers to act as combat commanders. Showing the flag was deemed to be more important than showing military prowess. In marked contrast to warfighting practice, peacekeeping involved the use of minimum force and placed a premium on mediation and qualities such as a predisposition for compromise and a disinclination to rock the boat. UN operational procedures also usually called for the rapid escalation of low-level incidents up to the highest diplomatic level for resolution. An associated rule of thumb held the media to be the main weapon of the combat soldier turned diplomat, policeman, and civil affairs bureaucrat. The very nature of peacekeeping, then, which equates more to internal security and police duty than warfighting, tended to militate against the enhancement of an army's professional operational capability.

Ironically, in both 1970 and 1990 Canadian troops were cast in the role of peacekeepers in their own country. During these two years, in fact, more Canadian troops deployed for peacekeeping-type duties at home than abroad; over 10,000 during the 1970 October Crisis in Quebec and most of a brigade during the Oka Crisis with the Mohawks twenty years later. Many Canadians who watched the Oka stand-off nightly on television were highly impressed by the restrained manner in which their soldiers, in the face of great provocation, defused the situation. Indeed, the operation appeared to confirm the value of peacekeeping experience in aid of the civil power and internal security duty. The old army argument that trained soldiers could be turned into peacekeepers was also shown to be true. What was not so obvious, however, was the weight of historical evidence indicating that trained peacekeepers could not so easily be transformed into soldiers. For example, after three to four years of the internal security equivalent in the Indian Empire, which placed a premium on inspections and smartness to overawe local inhabitants, the average British unit returned almost utterly useless for continental warfare. Similarly, the Tsarist army suffered an irreversible professional degradation and disintegration while engaged in internal security duties during the critical years 1905-14. The Argentine army was better trained to beat its own people than its British opponents during the Falklands campaign in 1982. The salvation of Canadian military professionalism is not, therefore, likely to be found in the tragedy and turmoil of Bosnia, which despite all talk of aggressive peace enforcement remains a commitment of a far different order and form than warfighting. To confirm this is so, one merely has to refer to casualty counts in Canadian military history.

NOTES

1. New commands announced in 1965 were maritime, mobile, air defence, air transport, training, and materiel.
2. Gordon A. Craig, "Prophets," *Times Literary Supplement*, 13 December 1996, 37.
3. The combined air-sea transportable (CAST) brigade assigned to the NATO northern flank between 1968 and 1987 was also a military sham of major proportions.

Chapter Six

Toward a Hegelian Army

Failure to preserve the institutional and educational basis of the military professionalism sustained by Simonds and his protégés inevitably led to the denouement of the Canadian army in Somalia. Twenty years earlier Simonds had also expressed concern about the long-term effect of unification on discipline, which he categorized as being of a different order within each service. In the navy the crew of a ship were literally all in the same boat with little individual choice but to sail where the captain sailed. The same could be said for air crew, but because they were sharply differentiated from ground crew, they had little responsibility for the command and administration of personnel comprising the latter. The nature of land warfare, on the other hand, presented almost every fighter in a unit with the opportunity to drop out of battle. In aid of civil power duty such as Oka, moreover, any failure of discipline could produce disastrous consequences as every soldier carried a lethal weapon. With some prescience, Simonds further stressed the "unwritten line between legitimate acts and orders from higher authority which were to be obeyed, and orders contrary to human compassion leading to the commission of war crimes." In all cases, he insisted, the inspiration of loyalty, discipline, and morale within a military body depended largely upon its corps of officers. Plainly put, there could be no such thing as a good unit without a good commanding officer. He might also have added that there could be no good army without good generals.

The Canadian army that served in Somalia went a long way toward confirming Simonds' worst fears. That it was a *new* army, the social laboratory spawn of a peacetime military establishment that had forgotten more than had been learned in two world wars and the prewar period, needs also to be stressed. Constant buffeting by the cross-currents of unification, bilingualism, and peacekeeping had, in addition to the overt civilianization of NDHQ, seriously eroded the professional foundations of army educational and training establishments set up after the war. With little depth and less corporate knowledge, there thus developed a widespread tendency to believe that the only kind of experience that counted was one's own, preferably gained in officially approved usually highly bureaucratized slots. This belief in turn encouraged an underlying anti-intellectualism. Unlike US Army posts, Canadian army bases boasted no military bookstores, which militated against getting new ideas from old sources, thus helping to leave the army as a whole more prone to fads. Whether one was engaged in endless army tactical doctrine board deliberations or actively training troops, the Canadian military system called for only the most superficial study of the profession of arms. This no doubt explained why Colonel Labbé, who commanded but 850-odd soldiers and a supply ship in Somalia, could so easily believe he was a practitioner of the operational art on a par with Montgomery and Eisenhower. The amateurish Canadian attempt to lead a multinational UN relief expedition to Rwanda in 1996 also saw Canada's army commander race off to Africa to lead a reconnaissance patrol when he should properly have remained at the seat of government. In many ways, the times seemed as militarily out of joint as they reputedly were after the purge of the Red Army in 1937:

> ...the men who followed...lacked any intellectual curiosity simply because they disposed no intellect, either singly or as a group. They mouthed slogans but understood nothing of principles...they were martial in a swaggering sense without the least grasp of the professionalism necessary to the military.[1]

What, then, can be done with the Canadian military to restore military professionalism and thus better serve Canada's citizens? There are good grounds for believing that, in contrast to Soviet experience, a purging of Canadian army brass could prove salutary. The gentler 1937 purging of general officers in the British Army, for example, carried that force kicking and screaming into the twentieth century. In the Canadian case, however, there are venal as well as professional military grounds for considering such action. The release of the army's top officer, Lieutenant-General Armand Roy, for claiming expenses to which he was not entitled should have been accompanied by charges being laid, as they most certainly would have been had he held a rank other than general officer. Roy not only disgraced his high office but arrogantly abused the trust so mistakenly placed in him. Judging from allegations made in the potboiler *Tarnished Brass*, which apparently has not drawn any defamation suits, it would be foolish to assume that the same flawed promotion pattern that spawned Roy did not create others in his image. As the Duke of Wellington once supposedly observed, "The cream always rises to the top, but so does the scum." One way to rid the general officer corps of the latter, would be to give the DM back the separate staff that once enabled him to act as an effective financial watchdog of the defence department.

There is also something unethical about a former CDS, Paul Manson, using his influence to sell helicopters to an organization he once headed. All generals should be prohibited from doing business with the defence department for a period of three years after retirement. In a related vein, retired generals of the regular army should also be forbidden by statute to assume honorary appointments as colonels of regular regiments. These positions should instead be filled by civilians to further good civil-military relations, while retired regulars should be left free to assume honorary appointments with militia units in order to foster better relations there. Those politically correct generals who prohibited Canadian soldiers from consuming any alcohol at all while serving overseas should also be cashiered for poor judgement, if not moral cowardice. The problem of troops drinking to excess has always reflected poor leadership rather than the evil of alcohol itself, which if consumed in moderation can indeed be a powerful morale booster. Imposing prohibition, which is almost impossible to enforce anyway, may additionally drive troops to consume poisonous home brew or dangerous drugs that have no equivalent redeeming features. Here again, taking the politically correct low road is as likely to prove as counter-productive as the search for a peacetime military relevancy unrelated to warfighting. It is hard to imagine, frankly, that the Canadian people really want a kinder, gentler army of plaster saints as opposed to a quietly tough, highly disciplined, professional force that knows its business.

Getting to this state, however, will require more critical analysis than is likely to flow from an improved public relations machine. The myth that reform can be effected through the redesign of NDHQ, by adjusting dotted and solid lines on organizational charts, has also for more than a generation been shown to be an utterly futile exercise. Neither is any one person on a white horse likely to have better answers. What is required is a collective effort by a smarter officer corps imbued with what used to be called character. The institutional education and promotion examination system cultivated during the Simonds era went a long way toward accomplishing just this. The purely theoretical, watered-down examinations and career courses that gradually replaced the examinations under Simonds were never up to the same intellectual standards or administered with the same sense of seriousness. The army adage that any fool can pass a course remains only too true. As well, the course approach to army education, besides taking men away from units, often militates against self-directed learning and the inculcation of that inquisitiveness that Clausewitz held to be more important than creativity in a leader. An officer should not be allowed to shirk personal responsibility for keeping himself or herself informed about developments in the profession of arms. This is why a university degree is never enough in itself, especially if graduates prefer running to reading an old book to get a new idea. Then, too, as the Japanese car manufacturer, Soichiro Honda, remarked, "The trouble with academics is that they only use their heads." Soldiers have to be able to translate theory into practice, which means that they have to be students of war as well as practical people. The higher they rise, the more important it is to be the humble student.

From a strategic perspective, Canadians have never historically been able to muster sufficient strength to defend their own country. Given the sheer size of the Canadian land mass, this most basic but often overlooked strategic reality is unlikely to change in the future.

Canada, therefore, has to have allies in order to ensure its national security. But, as always, alliances come with price tags rarely quoted in terms of peacekeeping, constabulary, or lightly armed gendarmerie forces. On the one hand, proportionate costs and burdens must be shared. On the other, expenditures may also have to be made just to prevent an ally from encroaching on national territory. Operating within large alliance structures has additionally meant that the three environmental branches of the Canadian armed forces, because of differences related to the effective employment of land, sea, and air, cannot practically fight together. To employ Canadian air forces just to support a Canadian army, for example, would be a wasteful use of assets that could be more effectively employed centralized in alliance with other air forces. A similar case can be made for the use of naval resources. The conclusion is therefore inescapable that the development of a purely national *military* strategy is likely to remain well beyond Canadian reach. But this does not mean that Canada cannot retain control of her armed forces to gain a greater voice in the diplomatic councils of power as in the Great War. The principle of fighting the army as an entity has long been established and so long as the same principle can be applied to the navy and air force there should be no problem with placing Canadian forces under separate allied operational control. In the case of a major conflict, sheer organizational demands will also probably require splitting the defence department into separate service cabinet ministries as in the Second World War.

Another reality worth recognizing is that the Canadian people from the South African War through the Great War, the Second World War, and the Korean War have consistently called for getting their troops into action overseas. On three different occasions they also clamoured for conscription, which means that they are probably quite prepared to do the same again should another occasion warrant. Of course, no one can foretell the future, which is why most nations have tended to maintain military establishments in peacetime as a form of security insurance. The cost of this insurance for a Canadian family of four runs to approximately $1500.00 per year. In 1991-92, however, defence estimates called for devoting 57 per cent of the budget to personnel related costs, mainly to sustain the regular force establishment. By another rough measure, of approximately ten billion dollars devoted to defence in 1997, nine billion went to field about 65,000 regulars and one billion to field around 30,000 primary reservists, which worked out to $138,461 per regular and $33,333 per reservist. But the question must be asked in respect of soldiers, who is really the more cost effective? From all indications, regular force personnel may be close to pricing themselves out of existence, especially as all too many of them constitute little more than an army of clerks. A militia soldier, conversely, can be brought up to regular force combat standard in three months. Just because a person in is the regular force, therefore, does not necessarily make him or her an expert in the profession of arms. In fact, on the basis of actual field time and exercise attendance, there is good reason to believe that from the top down there are a lot of soldiers in the regular army who are not professional or expert enough in warfighting to deserve the comparatively lucrative pay and allowances they are receiving. On the other hand, the only reason that certain militia units have not been able to recruit adequate numbers over the years is simply because they have not received enough financial and materiel support to offer sufficient enticements. One can make more money bagging groceries.

Apart from considerations of quality, it should also be evident that the regular army has long passed the point of being able to fulfill its commitments quantitatively. In Bosnia, for example, the militia contributed twenty per cent of Canada's force. Clearly, as never before since the Second World War, the regular army needs to be buttressed by the militia. But this buttressing should take the form of using the militia as a pool of all ranks talent rather than as merely a basic training reinforcement pool for filling regular army personnel shortfalls. A positive step in ensuring such an across-the-board revitalization would be to reassign the militia its traditional role as the mobilization base of a national army comprising several corps.[2] Bearing in mind that the militia originally lost this role during the era of massive retaliation when "forces in being" were the vogue, it would seem to make sense to reintroduce it during a period where conventional warfighting doctrine has for some time, at least since the introduction of NATO flexible response in 1967, increasingly held sway. As the militia continues to constitute the army's most important institutional link with Canadian society, this would also be a major step toward returning the army to the Canadian people (more than 700,000 of whom served in the army in the Second World War). Naturally, to accomplish this initiative, the militia will have to be expanded and allocated a much larger slice of the army budget, say, double what it receives now. At the same time, the regular army will have to be assigned, as it once was, the primary role of training the militia, *which role, in any case, it would have to assume in a major war.* With four times the number of personnel paraded by the permanent force during the interwar years and equipped with simulation technology that the latter did not have, the regular army of today should be able to do a much better job of offering challenging and rewarding training to Canadian citizens in the profession of arms. Perhaps, then, too, any fit able-bodied citizen who wished to serve his or her country in uniform could be allowed to do so, under mutually acceptable terms, without having to go on an extended regular force waiting list.

On being asked what type of army he wished for Britain, Lord Haldane, the newly appointed secretary for war in 1905, apparently replied that he preferred "a Hegelian army." As Haldane was an avid student of German philosophy and an ardent advocate of national education, what he probably had in mind was an enlightened force driven mainly by ideas dialectically honed through rigorous analysis and candid rational debate. Being largely responsible for the introduction of the British General Staff system, a new reserve territorial army, and the dispatch of the best equipped and trained army ever to leave Britain's shores, he was also as good as his word. In the space age where ideas and knowledge appear to be transcending and radically transforming the globe, substantial benefits could no doubt flow from the adoption of a roughly similar approach. To be sure, the encouragement of creativity, inquisitiveness, and *self-criticism* within the Canadian army would certainly lend it much of the renewed strength it so desperately needs. Canadian military history would also provide a ready fountainhead of knowledge for the sustainment of such an approach and underscore the requirement to retain a warfighting focus. As George Grant so perceptively posited, more-over, "if one cannot be sure about the answer to the most important questions, then tradition is the best basis for the practical life."

The Canadian army has a long and honourable tradition of which all Canadians can be justly proud. A major step toward returning it to its roots would be to concentrate on producing a proper army General Staff with troops to support both a corporate knowledge base and one unbroken command chain of mutual respect and trust running from top to bottom throughout the army. In short, the army high command should be one with its troop formations in the field. In order to produce the best possible officers to assume such duty, the army staff college should again become a true nursery of the General Staff on the leading edge of advanced military studies and research. Highly trained college graduates would, in turn, be able to train an expanded militia and participate in either capital or joint staff plan-ning as required, but they would also be there to facilitate the mobi-lization and fielding of a larger army for war.[3] The future of the Canadian army might then look bright indeed. Neither will this lament have been in vain. *Sunt itinera per silvam.*[4]"

NOTES

1. John Erickson, *The Road to Stalingrad* (London: Weidenfeld and Nicolson, 1975), 7.
2. Although no major or minor wars loom on the horizon at pre-sent, the threat of Quebec secession does, and Canada may well have to defend its interests by providing a brigade to protect Labrador and another two to retain the originally British territo-ry of Ungava in northern Quebec. These would also have to be militia brigades as a regular brigade would be better retained as an immediate reaction force.
3. To ensure that some generals at least remain with field forces, command of brigades should not, as recently proposed, be given to colonels, whose grade between the normal wartime command progression of lieutenant-colonel to brigadier has traditionally been used as a staff rank.
4. "There are roads through the forest."

Further Reading

Appendix 2, "Whatever Happened to the National Defence College of Canada?", is a paper written by the author in 1993 on the National Defence College (NDC) of Canada, then the top educational institution for senior and general officers of the Canadian Forces. Prepared while the author was still a serving officer on the NDC staff, the paper uses primary source material to trace the erosion of the intellectual foundations of military professionalism and the rise of the bureaucratic ethic within this institution. Although the NDC was closed shortly after *Maclean's* and *Canadian Business* magazines published editorials drawing attention to the paper, this original work should continue to serve as a valuable benchmark against which to measure any future courses designed to train and educate the highest ranking officers of the Canadian Forces. In short, if such courses ever again approach the professional mediocrity of NDC as recorded in 1993, they too should be terminated.

For a broader and more profound understanding of Canadian military professionalism, *Canada's Defence: Perspectives on Policy in the Twentieth Century* (Toronto: Copp Clark Pitman, 1993), edited by B.D. Hunt and R.G. Haycock, provides a good starting point. *Canadian Military History: Selected Readings* (Toronto: Copp Clark Pitman, 1993), edited by Marc Milner, provides further insights into the operational as well as institutional aspects of the armed services. Although the indispensable guide to the early development of Canadian military professionalism remains Steve Harris' classic *Canadian Brass: The Making of a Professional Army 1860-1939* (Toronto: University Press, 1988), Desmond Morton's *A Military History of*

Canada (Edmonton: Hurtig, 1985) and G.F.G. Stanley's *Canada's Soldiers: The Military History of an Unmilitary People* (Toronto: Macmillan, 1974) can still be read with profit. On the Canadian experience in the Boer War the definitive work appears to be Carmen Miller's *Painting the World Red: Canada and the South African War, 1899-1902* (Toronto: University Press, 1992).

Among the seminal works of the past that continue to stand in a league of their own are: Richard A. Preston's *Canada and 'Imperial' Defence: A Study of the Origins of the British Commonwealth's Defense Organization, 1867-1919* (Durham, NC: Duke University Press, 1967); C.P. Stacey's *Canada and the Age of Conflict: A History of Canadian External Policies, Volume 1: 1867-1921* (Toronto: Macmillan, 1977); Desmond Morton's *A Peculiar Kind of Politics: Canada's Overseas Ministry in the First World War* (Toronto: University Press, 1982); and A.M.J. Hyatt's *General Sir Arthur Currie: A Military Biography* (Toronto: University Press, 1987). The more recent additions of *Surviving Trench Warfare: Technology and the Canadian Corps, 1914-1918* (Toronto: University Press, 1992) by Bill Rawling, *When Your Number's Up: The Canadian Soldier in the First World War* (Toronto: Random House, 1993) by Desmond Morton, and *Shock Army of the British Empire: The Canadian Corps in the Last Hundred Days of the Great War* (Westport, CT: Praeger, 1997) by Shane B. Schreiber have further served to increase our general understanding of the great events recorded in the *Official History of the Canadian Army in the First World War: Canadian Expeditionary Force, 1914-1919* (Ottawa: Queen's Printer, 1962) by G.W.L. Nicholson.

Among works on the Second World War pride of place goes to G.W.L. Nicholson's *Official History of the Canadian Army in the Second World War*, Volume II, *The Canadians in Italy 1943-1945* (Ottawa: Queen's Printer, 1962) and Colonel C.P. Stacey's *Official History of the Canadian Army in the Second World War*, Volume I, *Six Years of War: The Army in Canada, Britain and the Pacific* (Ottawa: Queen's Printer, 1966), *Official History of the Canadian Army in the Second World*, Volume III, *The Victory Campaign: The Operations in North West Europe, 1944-1945* (Ottawa: Queen's Printer, 1966), *Arms, Men and Governments: The War Policies of Canada 1939-1945* (Ottawa:

Information Canada, 1974), and *Canada and the Age of Conflict; A History of Canadian External Policies, Vol. 2: 1921-1948, The Mackenzie King Era* (Toronto: University Press, 1981). The complexity of ground force operations and the associated need for training and warfighting knowledge is also the focus of the author's own *The Canadian Army and the Normandy Campaign: A Study of Failure in High Command* (New York: Praeger, 1991) [republished as *Failure in High Command: The Canadian Army and the Normandy Campaign* (Ottawa: Golden Dog, 1995], Reginald H. Roy's *1944: The Canadians in Normandy*, Canadian War Museum Historical Publication No. 19 (Ottawa: Macmillan, 1984), and Terry Copp and Robert Vogel's *Maple Leaf Route* (Alma: Maple Leaf Route, 1983-1988) series. In roughly the same genre are works by Jeffrey Williams, *The Long Left Flank: The Hard Fought Way to the Reich, 1944-1945* (Toronto: Stoddart, 1988), Daniel G. Dancocks, *The D-Day Dodgers: The Canadians in Italy, 1943-1945* (Toronto: McClelland and Stewart, 1991), Brian Loring Villa, *Unauthorised Action: Mountbatten and the Dieppe Raid* (Toronto: Oxford University Press, 1990), and W. Denis and Shelagh Whitaker, *Dieppe: Tragedy to Triumph* (Toronto: McGraw-Hill Ryerson, 1992), *Tug of War: The Canadian Victory that Opened Antwerp* (Toronto: Stoddart, 1984), and *Rhineland: The Battle to End the War* (Toronto: Stoddart, 1989). A closer look at Canadian field commanders is additionally offered in J.L. Granatstein, *The Generals: The Canadian Army's Commanders in the Second World War* (Toronto: Stoddart, 1993) and Dominick Graham, *The Price of Command: A Biography of General Guy Simonds* (Toronto: Stoddart, 1993).

Defence policy and military institutional development between the Great War and the present day is brilliantly covered by James Eayrs in his uniquely analytical series *In Defence of Canada,* 4 volumes (Toronto: University Press, 1964-85), which remain books to be studied as much as read. While the best work on the army in Korea is still Herbert Fairlie Wood's *Strange Battleground: The Operations in Korea and their Effects on the Defence Policy of Canada* (Ottawa: Queen's Printer, 1966), J.L. Granatstein and David J. Bercuson in *War and Peacekeeping: From South Africa to the Gulf—Canada's Limited Wars* (Toronto: Key Porter, l992) provide a longer perspective on Canadian military involvement in small wars overseas. Bercuson's *True Patriot: The Life of Brooke Claxton 1898-1960* (Toronto: University Press, 1993) also sheds light on Canadian military policy from the outbreak of the Korean Conflict and the Cold War. A glimpse of the national defence bureaucracy from a more contemporary standpoint can additionally be found in Douglas L. Bland, *The Administration of Defence Policy in Canada, 1947 to 1985* (Kingston: R.P. Frye, 1987) and his *Chiefs of Defence: Government and Unified Command of the Canadian Armed Forces* (Toronto: Brown Books, 1995). In a related vein, the prescient "Commentary and Observations" by Lt. Gen. G.G. Simonds in *The Canadian Military: A Profile* (Toronto: Copp Clark, 1972) edited by Hector J. Massey still appear remarkably relevant. For the best treatment of defence issues during the Trudeau years one should consult J.L. Granatstein and R. Bothwell, *Pirouette: Pierre Trudeau and Canadian Foreign Policy* (Toronto: University Press, 1990). On peacekeeping, the only scholarly work produced since 1962 by a Canadian peacekeeper is James H. Allan's *Peacekeeping: Outspoken Observations by a Field Officer* (Westport: Praeger, 1996). On Somalia, Bercuson's *Significant Incident: Canada's Army, the Airborne, and the Murder in Somalia* (Toronto: McClelland and Stewart, 1996) is likely to be but the first of a long scholarly line.

As for the effect of war upon Canadian society, there is no finer work than John Herd Thompson's formidable *The Harvests of War: The Prairie West, 1914-1918* (Toronto: McClelland and Stewart, 1978), which definitely deserves to be read by more Canadians. Also invaluable for a portrait of Canadian society at war are *Ontario and the First World War, 1914-1918: A Collection of Documents* (Toronto: University Press, 1977) edited by Barbara M. Wilson and *Canada 1896-1921: A Nation Transformed* (Toronto: McClelland and Stewart, 1974) by Robert Craig Brown and Ramsay Cook. Unfortunately, there is no equivalent study of the Canadian home front during the Second World War to match that of Herd's *Harvests of War*.

Appendices

The Organization of the Canadian Corps in 1917

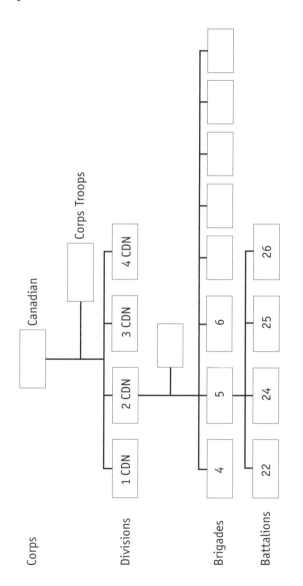

By 1918 a brigade of engineers had been added to each division, and machine-gunners were grouped in a battalion for each division as well.

Appendix 2

Whatever Happened to the National Defence College of Canada?

The National Defence College (NDC) of Canada located behind the stone grey walls of historic Fort Frontenac in Kingston, Ontario, remains like the fabulous griffin on its crest largely a mystery to the Canadian taxpayer who foots its bills. The permanent staff of the NDC proper, hardly more than a corporal's guard, actually belies its cost: a commandant in the rank of major-general, a vice-commandant of ambassadorial status seconded from the Department of External Affairs, a colonel coordinator, two directors of colonel rank, a civilian director, a lieutenant-colonel planner, two majors, a captain, and roughly eleven non-commissioned and civilian support personnel. The harsh reality, however, is that annual expenditure for the approximately 45 senior officers and civilians normally selected to attend the 44-week NDC course usually exceeds $4,500,000.00, or $100,000.00 each, in addition to salaries and contributions. That NDC military and secretarial support salaries amounted to some $2,671,760.00 in 1991-92[1] further attests to the relatively enormous, and possibly extravagant, sums spent on the operation of this institution.

The officially sanctioned mandate of the NDC is to prepare "senior military officers and civilian officials for appointments at the highest levels in their respective organizations by enabling them to study together in a mature environment those aspects of national and international affairs which determine or significantly affect the security of Canada."[2] For officers of the Canadian Forces attendance on the NDC course supposedly represents the "ultimate step" in their career development. The typical course, intentionally loaded with less than half military, usually includes 14 Canadian Forces' officers of either colonel or brigadier-general rank, a like number of federal and provincial civil servants of "equivalent" status, three police superintendents, three to six members from the private sector or academia, and eight foreign military or diplomatic senior officers. The curriculum comprises five terms: a one week orientation; three months of Canadian studies; a month of Canadian-American studies; five months of international studies; and one month of reviewing the year's work and postulating future Canadian national policies in an "exercise" termed "Whither Canada." In addition to participating in discussion and study groups, course members receive over 100 lectures from ambassadors, journalists, academics, captains of industry, and government officials on current issues of the day. They are also required to produce a 5,000 to 10,000-word paper on a subject not in their area of expertise.

What really makes the NDC unique, however, is its extensive travel program that for decades has carried course members and their wives to all provinces and continents. No other nation's defence college travels so widely as does the NDC. Three major field visits are regularly conducted within Canada: the longest to the North and West; one to Atlantic Canada; and one within Central Canada, mainly Ontario and Quebec. A two-week trip to the USA is also generally slated for the Canadian-American studies term. Three international tours, some of four or five weeks duration, are additionally made to South America, Asia, Africa, and Europe. The 1977-78 course travelled for 121 days, 31 of them consecutively in Asia and Africa, logging about 226 flying hours and 78,050 air miles. They visited 17 countries overseas, including Japan, China, India, Zambia, Nigeria, Tunisia, Trinidad and Tobago, Brazil, Jamaica, Germany, Egypt, Israel, Greece, Yugoslavia, Portugal, Norway, and Belgium. The 1983-84 course similarly travelled over 100 days, visiting Japan, Singapore, Indonesia, Australia, Cuba, Brazil, Costa Rica, Jordan, Israel, Czechoslovakia, Belgium, France, and Germany. By 1988 the NDC travelled for 41 days in Canada and 89 days visiting as many as 20 countries.[3] One former student and commandant who described himself as an "an aging military bureaucrat" in 1979 managed to visit more than 60 countries while he was with the NDC.[4] Not surprisingly, the cost of tours and visits in 1990-1991 totalled some $1,820,713.00.

For many years defenders of the NDC have argued that the course is of "incalculable" benefit to Canada and therefore worth every penny spent on it. A sprinkling have even written articles to this effect, some bearing such prepossessing titles as "The Best Year of Your Life"[5] and "Role of a Lifetime."[6] In the opinion of this group, the NDC more than adequately fulfills its mandate by considering national security in the broadest possible context. The diversity of course membership is intended to provide a favourable milieu for the cross-pollination of ideas. Collegial discussion, exposure to distinguished speakers, personal study, and travel serve in turn to produce "better people"[7] more able to appreciate the complexities of national and international issues affecting Canadian national security. Defenders further extol the NDC course as a uniquely individual experience. Several even suggest that course members can, like Saul on the road to Damascus, undergo spiritual rebirth. One senior public servant upon graduation quit his job and joined a religious seminary. One transmogrified commandant even renounced the profession of arms to crusade for world peace and disarmament.

Since not everyone with knowledge of the NDC can be considered a supporter, however, the case for the defence of "Canada's Best Kept Secret"[8] bears closer scrutiny. Does it really, as recently asserted by one Vice-Chief of the Defence Staff, provide the successful graduate with the military equivalent of an M.A. or Ph. D.? Or is it more akin to the Canadian Centre for Management Development[9] and Queen's University Program for Public Executives,[10] which both employ a similar guest lecture and "permanent syndicate" pedagogical approach to cover fundamentally the same political, economic, social, and environmental issues as the NDC? In many ways, the NDC course approximates a self-fulfilling prophecy; course members are told from the start that they are important people, and the undiscerning among them readily assume this to be true. An NDC graduate association that meets annually reinforces the perception that they attended an important course. The social advantages of having attended the NDC have also been elucidated as follows:

81

Don't you wish you could launch a cocktail party conversation with something more exotic than your current T[reasury] B[oard] submission or a learned commentary on the Cab[inet]Doc[ument] format and its relevance to decision making? Wouldn't you rather answer the usual, "Hi! What's new? I haven't seen you for a while," with something like "Well, having just sampled the long-range policy concerns of countries with over two thirds of the world's population, I think we will be in real difficulty if we can't pull together a coherent national strategy." Such an opener could raise eyebrows, provoke offers of another drink before the bar closes, or at least a polite, wondering, wait for your next sentence.

For about forty people every year [NDC graduates], such conversations are assured for life.[11]

Canada's NDC can, of course, lay claim to being the third oldest national defence college in the world. Although established late in 1947, its roots go back to the Imperial Defence College (IDC), the first such institution attended by Canadians. The IDC opened in January 1927 charged with "training...a body of officers and civilian officials in the broadest aspects of Imperial Strategy." Under the direction of its first commandant, the brilliant sailor-scholar Vice-Admiral Sir Herbert Richmond, who played a key role in its foundation, the college nurtured an intellectual approach to the study of war. A tireless advocate of closer interservice cooperation and the need for a common doctrine, Richmond encouraged students to consider British imperial defence not only from a military perspective, but in the wider context of the political, diplomatic, economic, social, industrial, logistic, and financial dimensions affecting a thalassic empire. Included among those aspects examined were the higher direction of war, the defence of bases and protection of trade, the use of convoys and methods of controlling or destroying enemy commerce, the treatment of neutral states, the geography of manufactures and command of markets, the influence of public opinion on the conduct of operations, the definition of commanders' responsibilities in combined operations, and various country or area studies. Occasionally, the IDC considered concrete problems referred to it by service chiefs of staff. By studying national strength in all its forms, the IDC sought to produce military officers who were more broad-minded and civilians who were better informed on defence matters.[12]

Richmond was one of the brightest commandants ever to be appointed to IDC. A former captain of HMS *Dreadnought*, naval historian of note, and founder of the *Naval Review*, he ranked among the foremost military theorists of his time. He was also an experienced educator who understood that an open atmosphere which permitted free discussion and debate facilitated learning. At the same time, Richmond was a practical enough sailor to realize that individuals could not learn from lectures alone; they had to learn by doing and reasoning things out for themselves. Accordingly, he each year conducted around a dozen exercises, some over a week long, in which students representing ministers and service appointments analyzed crises situations, prepared staff papers, organized responses, and generally played out their roles, which were, in turn, later criticized in open forum. Some of these exercises, which continued up to the Second World War, presaged events in that conflict to a remarkable degree, most notably that Singapore would be attacked from the landward side and that Hong Kong could not be defended militarily. Significantly, though day visits were made to defence establishments and industrial parks, it was not until 1929 that the IDC proceeded abroad on tour. The occasion was a Great War battlefield tour approved as an exceptional measure, since "it did not fall within the normal scope of the curriculum of the College."[13]

In 1939 the IDC closed for the duration of the war. It reopened in 1946, the same year the NDC was conceived, under the commandantship of General Sir William Slim with Canadian Lieutenant-General Guy Simonds, the only non-British officer ever to serve on the IDC staff, as his Senior Directing Staff (Army). Simonds, Canada's most outstanding higher field commander in the Second World War, was eventually named second commandant of the NDC to enable him "to acclimatize himself in Canada." The honour of being the first commandant fell to a British officer and 1939 IDC graduate, Major-General J.F.M. "Jock" Whitely, who also served with Simonds on the Directing Staff of IDC during the period 1946-47.

Whitely, loaned from the IDC, advised the Canadian government on the establishment of the NDC and strongly urged the adoption of the IDC model. Significantly, his espoused views ran counter to those of the government's own principal military advisers, General Charles Foulkes, Chief of the General Staff, and Air Vice Marshal W.A. "Wilf" Curtis, Acting Chief of the Air Staff, who both advocated the formation of a Canadian Joint Services Staff College. As originally envisaged by Foulkes, such a college would "preserve the techniques, mutual understanding, high spirit of joint service cooperation, interdependence of equipment design, and common doctrine which were major factors in the Allied victory."[14]

In the event, the IDC concept received its greatest support from the civilian sector of the public service. The Secretary to the Cabinet, Mr. A.D.P. Heeney, and the Under Secretary of State for External Affairs, Mr. L. B. Pearson, each sided with Whitely. Pearson thought that a Joint Services Staff College would be too military and proposed instead an institution wherein military minds could be steeped in civilian modes of analysis. He additionally suggested that if the External Affairs department and other government agencies were to be expected to financially assist and participate, the course of study would have to increase the vocational usefulness of their respective candidates. On 9 July 1947 Heeney and Pearson submitted a paper to this effect to the Chiefs of Staff Committee, calling for the establishment of the NDC by name. Though opposed by both Foulkes and Curtis, the Pearson-Heeney paper received the approval of two civilian members, the Chairman of the Defence Research Board, Dr. O.M. Solandt, and the Deputy Minister of National Defence, Mr. W.G. Mills. The deciding vote in favour of the NDC was cast by the Chief of the Naval Staff, Vice Admiral H.E. "Rastus" Reid, who was less enamoured of a military-civil service college than fearful of an integrated Joint Services Staff College. His action was probably a parting shot in a long interservice vendetta.[15]

The NDC opened in December 1947 and commenced its first course on 5 January 1948. At the opening ceremonies, The Honourable Brooke Claxton, Minister of National Defence, observed that it represented "the first effort to organize in this country an institution for the advanced study of war and security problems." The course focussed on war as a totality and defence as a single study, for in Claxton's words, "the recent war [proved that] the real power that endures and wins is a tight and undivided combination of the industrial and the military, the scientific and the strategic, finance and trained manpower, the laboratory and the parade ground, the railway yards and reinforcement camp, the sheltered man of ideas and the rugged man of arms."[16] As has been shown, of course, this essentially reflected the approach of the pre-war IDC. The major difference was that the NDC began to travel extensively, probably in imitation of the postwar IDC, which at General Slim's instigation commenced visits to various national capitals "on the grounds that students would learn more ... in an organized *short* visit than in a much *lengthier* period of study at home."[17] But whereas IDC travel abroad started slowly and rarely exceeded a month, NDC travel overseas was often triple that amount. Visits to Africa and Asia, which began as an adjunct to the NDC European tour, became more and more extended until two trips proved necessary. During 1967-68 the course spent 75 days out of Canada, and a total of 89 out of 234 working days on tour. By 1988 travel consumed more than half of total course time.[18] Indeed, travel so pervaded the NDC course that its curriculum gradually began to orientate around field trips.[19]

As the urgencies of the 1950s devolved into Cold War strategic stalemate, both the NDC and the IDC acquired reputations for being centres of leisure rather than centres of learning. As described by one authority, the NDC was "an educational disaster, lacking curriculum and competence, and exposing its students to a few defence problems with an intensity appropriate to a gentleman's club."[20] To satisfy a growing criticism that its course was insufficiently demanding, the IDC in 1967 invited students to write voluntary theses. The 1970 reforms of the The Honourable Alistair Buchan, which included the compulsory requirement for an 8,000 to 10,000-word student paper, specifically aimed at raising the intellectual standards of the IDC, which also saw its name changed to the Royal College of Defence Studies (RCDS).[21] Ironically, the NDC had meanwhile been investigated by an IDC graduate, Dr. Frank Milligan, Assistant Director of the Canada Council. In September 1967 Prime Minister Pearson requested Milligan to undertake a reassessment of the role of the NDC, including its value to the public service as a whole and its relationship to other senior personnel training and management courses available or planned for the future. Shortly after the study commenced in October 1967, the Cabinet endorsed an earlier recommendation of the Cabinet Committee on Priorities and Planning (CCPP) calling for the closure of the NDC as early as 1968 in order to save money. As this would have meant closing the NDC four or five months before Milligan rendered his evaluation, however, the Prime Minister and Cabinet agreed to delay their decision until after receipt of his report.

Milligan eventually submitted his report to Prime Minister Pierre Elliott Trudeau in January 1969. It concluded that as the NDC was out of touch with the domestic realities of Canadian life and "largely irrelevant to the task of senior officer development in the public service," it should be closed. Milligan proposed instead a system of public service training based on senior executive bilingual courses, exchanges among government departments, sabbatical leaves, and earlier overseas postings for selected individuals during the course of their careers. Looking to the future, he further suggested encouraging universities to develop "programs for public servants." In Milligan's view, the armed forces and external affairs department both laboured under a "tradition of isolation" from Canadian domestic affairs, which other civil departments and agencies involved with the NDC would have preferred to see as the primary focus of study. Though Milligan recognized that armed forces had a "continuing need for special training programs devoted to their professional role and the international context in which it is discharged," he also noted the pronounced tendency of the newly unified Canadian Forces to seek "a radical recasting of the[ir] peacetime role." He thus naturally assumed that the development of senior military officers required "opportunities and programs akin to those needed in the civil sector." On the recommendation of the Clerk of the Privy Council and Secretary to the Cabinet, Trudeau referred the Milligan report to the CCPP for further review.[22]

In June 1969 the Advisory Panel on the Executive Category, which included representation from the Privy Council Secretariat, Treasury Board, and the Public Service Commission, agreed that the NDC should be closed and proposed to make the appropriate recommendation to the CCPP. Up to this point the Department of National Defence (DND) offered no argument to the contrary. In fact, a minute on the copy of the Milligan Report forwarded to the Chief of the Defence Staff in March 1969 read, "I am sure you will be very pleased to see Mr. Milligan's report."[23] What apparently prompted this sentiment was Milligan's warm praise for the armed forces in their "search for new roles" and for the work of the DND Officer Development Board chaired by Major-General Roger Rowley, a distinguished combat veteran of the Second World War. Rowley's "Report of the Officer Development Board," which also appeared in March 1969, levelled several devastating broadsides against the NDC. The most serious was that NDC students were neither "examined, nor required to do research in depth." The board further observed that the college itself lacked a research staff. Noting that there was "no school in Canada teaching higher military operations," the Rowley report concluded that the NDC could not "be considered...to be a career course" for military officers. It recommended instead the establishment of an "Advanced Military Studies Course" and an interdepartmental "National Security Course" backed up by a Centre for Strategic Studies that would conduct continuous and comprehensive research in the strategic realm.[24]

The Advanced Military Studies Course (AMSC) proposed by the Rowley board in 1969 was definitely ahead of its time. In essence, it resembled the School for Advanced Military Studies (SAMS) established by the U.S. Army in 1983 to teach the operational level of war. The aim of the 44-week AMSC was to train roughly 40 lieutenant-colonels each year in "joint high-level military operations." The board argued that though Canada might "never herself raise forces of the size studied at the proposed AMSC, ...a senior officer [had to] understand high-level operations, totally master his profession and be able to give effective advice to government and to function effectively in a multinational high-level headquarters." The AMSC course called for eight weeks of studies in strategy covering the elements of national power, war as a phenomenon, contemporary strategic theories and problems, civil-military relations, and the impact of conflict on the international system. It also incorporated twenty-five weeks of advanced studies in military operations, management, and technology, including exercises, war games, and research activities, which concentrated on such aspects as the higher dimensions of warfighting, force structure, field deployment planning, the economics of defence, research and development, budgeting, decision making, and weapons effects. The board also envisioned the AMSC as eventually leading to a master of military science degree for suitably qualified candidates who successfully completed the appropriate examinations and theses.[25]

The seven-month National Security Course (NSC) advocated by the Rowley report was to be interdepartmental in terms of staff, students, and financing, but DND administered. As explained by the naval commandant-designate of the NDC, Brigadier-General S. Mathwin Davis, the concept behind the NSC derived from the formula: "NDC + More Military Content = *National Security Course*."[26] The stated aim of the NSC was to prepare officers in the rank of colonel, senior public servants, and civil sector representatives for higher levels of responsibility in the conduct of national affairs through the study of national security problems. The proposed curriculum included a refresher on current theory and practice in military operations and a short introduction to the government policy process. Seven main study blocks focussed on contemporary strategic problems, the role of Canada in international relations, internal domestic issues, Canadian foreign and defence policies, international law, area studies, and the use of crisis gaming as a research, educational, and testing tool. While the report recommended a lecture-seminar approach for this program, the presumption was that the number of guest lecturers would not be too large and that discussions would "be led by well-informed persons." Equally significant, the report stressed that NSC "travel on a mass basis should be reduced to the barest minumim"; while not ruling out "justifiable" visits, it envisaged most travel being conducted by "individuals or small groups in pursuit of their research problems." The proximity to the

NSC of an associated Centre for Strategic Studies, replete with a permanently resident staff of experts and computer-based comprehensive gaming facility, was intended to supplement the depth and breadth of NSC teaching, research, study, and practice.[27]

As might well be imagined, the relative coincidence of the Rowley and Milligan reports evoked a variety of responses from military and public service quarters. Initially, Chief of Personnel staff adopted the position that Milligan erred in equating higher military training with a "general system of executive training for officers in the Public Service."[28] Other comments forwarded to the Chief of the Defence Staff read: "Nothing in Mr. Milligan's report leads us to believe that we should deviate from the goal of the senior officer development scheme proposed in the O[fficer] D[evelopment] B[oard]."[29] At the same time, however, the NDC and Brigadier-General Davis began frantically to develop an approach of their own. The "main point," the NDC commandant-designate cautioned, "is our continued existence."[30] He clearly regarded the question of increased military course content as negotiable and ventured that the NDC "would probably opt for more non-military content—[NDHQ staff] for more military."[31] To Davis, there was a "measure of truth" in Milligan's assertions[32] and he suggested that the NDC observe certain "tenets" in the formulation of arguments, including basing all proposals on the Milligan report and never "discredit[ing] or contradict[ing] its major premises—which had already been accepted." The "appropriate interests" of the Public Service Commission, Treasury Board, and the Privy Council concerning bilingual and senior executive training and long-range planning were likewise to be recognized. The "undoubted lack of relevance of much of the course to many Public Servants and the need to improve this" was also to be admitted. The NDC could, in short, "be recast to meet any reasonable aim."[33] As for "the world wide sight-seeing for which NDC now seem[ed] notorious," Davis suggested that "perhaps if we demonstrate, in a proposed course outline, that there would be *less*, this might be adequate."[34]

Ultimately, the inveterate Davis, who served as commandant from 1969 to 1974 in the rank of rear admiral, proved to be the most bureaucratically adept at negotiating NDC course content. The natural hesitation of the Canadian Forces to close the NDC before the implementation of the Rowley board recommendations gave the institution a renewed lease on life. A proposed "modified NDC," which would have seen more military course content introduced, never truly materialized largely because of Davis's personal determination to demonstrate greater responsiveness and sensitivity to the needs of the public service. In January 1970 the Department of External Affairs, DND, Treasury Board, and the Public Service Commission collaborated in the development of a new program of studies. Trialled during the 1970-71 NDC year, it actually comprised two courses: a 16-week "Senior Executive Development Course" (SEDC) designed to increase civilian attendance and attended by all students; and a "Canadian, International and Military Studies Course" attended by 36 mainly military students. The following year the NDC reverted to a single course format, but with a greatly expanded focus on Canadian domestic and governmental issues. This fundamental shift directly reflected Davis's "personal inclinations and biases" that:

> ...if the college were to continue, there would have to be some demonstration of very significant change—some specific efforts that, if nothing else would give the appearance of constructive moves in the direction of "relevance". In short, it would have to become obvious that NDC was no longer primarily involved with national defence and the external manifestations of national security....[35]

In fact, adding domestic issues to the NDC curriculum, expanding course civil representation, and mounting a concerted publicity campaign, enabled Davis to win the support of powerful civil service mandarins. Mr. Michael Pitfield of the Privy Council, who formerly favoured the closure of NDC, supposedly in a change of heart personally persuaded Trudeau to allow the 1971-72 course to proceed. A subsequent Cabinet decision in May 1972 directed that the NDC would continue "until such time as the Executive Development Program in the Public Service is able to meet the needs for senior executive training of the military."[36]

In retrospect, it is easy to see that it was the impact of the

Milligan report rather than the Rowley board that set the course of
the NDC down to the present. Changes to the NDC course program
since 1972 have also continued to bear out the astute observation of
James Eayrs that "each year witnessed the progressive demilitariza-
tion of the curriculum."[37] To paraphrase Davis writing in 1974, the
NDC moved in this direction "by its own volition." He assessed the
college's primary purpose as the "enhancement of civil-military relations"; the second as "gaining a greater understanding for the very
wide range of topics political, social, environmental, economic, ethnic or cultural which are all comprehended as being part of 'a concern for national security.'"[38] Under Davis, the course nonetheless
remained "moderately involve[d] with the Defence category of the
Military Role"[39] and divided into four terms: Canadian studies dealing with mainly domestic issues; external affairs and international
relations; strategic and military studies; and the study of specific
problems related to planning, budgeting and systems analysis.[40] The
term titles employed in 1978 reflected a rough similarity—Canadian
Studies, The World Environment, Strategy and Canadian Defence,
and Whither Canada—but by 1981 the entire strategic term had
been eliminated.[41] A major review of the college undertaken in 1982,
largely in response to complaints about lack of course structure, poor
lecture flow, and a travel program haphazardly related to major
Canadian and international issues,[42] produced the current NDC
course alluded to earlier.

One might well question, however, whether the adoption of a
reworded mandate and revised course structure really amounted to
much more than a bureaucratic reordering of the window-dressing.
The essential focus of the NDC course remained what it had always
been since the Milligan report. That the NDC turned from defence to
domestic issues during this particular period should not be surprising, of course, as it was one "long, dark night of the spirit" for
Canada's recently unified forces:

> The Trudeau years had been extraordinarily difficult ones for the
> Canadian forces. Reviews had disrupted training and planning; bilin-
> gualism had strained goodwill within units; managers began to replace
> leaders; budget freezes and rampant inflation constrained the forces;

and lack of modern equipment helped destroy the regard in which allies had once held the Canadian military.[43]

While the Canadian Forces never fully recovered from the wounds inflicted upon them during these years, they did eventually manage to reverse several trends inimical to military professionalism. But the NDC remained a thing apart, continuing to reflect the unmilitary attitudes and priorities of the early Trudeau years. Closer examination of the prolix and pretentious reams of documentation written in defence of the college further reveals an institution bent upon rationalizing its curriculum for the sole purpose of prolonging its existence.

The NDC definition of national security remains a case in point. Whereas the prestigious Harvard periodical *International Security* defines the term "broadly to cover all matters pertinent to the use, threat, and control of force,"[44] the NDC employs the more circumlocutionary:

> The preservation of a way of life acceptable to the Canadian people and compatible with the needs and legitimate aspirations of others. It includes freedom from military attack or coercion, freedom from internal subversion and freedom from the erosion of the political, economic and social values which are essential to the quality of life in Canada.[45]

By this definition, which has no particular focus, almost anything could be included under the rubric national security. Given that "'security' means the absence of threats,"[46] of course, there is obviously more to the concept than the military dimension alone. Indeed, this reality has long been recognized by the more sagacious, even before the advent of Richmond's IDC.[47] Today threats to national security can arise from such things as overpopulation, trade disputes, and the indirect effects of "dwindling reserves of oil and the deterioration of the earth's biological systems."[48] The pressures of poverty, poor education, political oppression, ethnic rivalry, terrorism, industrial strife, crime, and pandemic disease all pose problems to individuals and states alike. Yet, while few deny that such international concerns ultimately affect national security, the list of factors indirectly bearing on this field of study is potentially infinite. It could even be argued that "potty training," since it may affect the development of generals, also constitutes a national security concern.[49] Common sense would therefore seem to suggest that some degree of focus is necessary.

By any reasonable measure, the military dimension still lies close to the core of national security and strategic analysis. As even admitted by the respected strategic critic, Ken Booth, regardless of how broad a grasp the student of security studies may have of human rights, environmental issues, and economic problems, "an understanding of defence [remains] essential."[50] Indeed, to quote a former Deputy Minister of the DND, C.R. Nixon, if "total security includes not only military defence but the environmental, health, shelter and other factors for a stress-free life...[then] military security is a prerequisite for these."[51] In the final analysis, then, Canadian senior officers should at least be well-versed in higher military affairs, for however great their awareness of political, economic, and social issues, they are unlikely to be permitted (probably wisely) to take specific action in these areas. As shown, however, the NDC preferred for reasons of expediency not to focus on military theory and grand strategy within the context of peripheral considerations, but chose instead to survey the peripheral aspects of national security giving only a passing glance to things military. In 1991 the course accepted briefings from the commanders of Air Command and Northern Region Headquarters that for their extreme superficiality could just as easily have been given to a group of visiting school teachers.[52] There is simply no depth to defence studies at the NDC. Consequently, despite its military trappings, the institution has nothing to offer the serious student of national security affairs. No sophisticated compendium of knowledge has ever been developed, nor have any authoritative publications dealing with national security, or any other related subject for that matter, ever been produced. There exists no NDC equivalent of the RCDS Seaford House Papers.[53] The so-called security analyses conducted by students on Canadian provinces(!) and various foreign countries are really nothing more than traditional "area studies" trussed up in vogue words.[54] No group of junior officers undergoing training at the Canadian Land Forces

93

Command and Staff College, collocated with the NDC in Fort Frontenac, would have any trouble excelling in their preparation.

In his work on American grand strategy, Paul Kennedy referred to the critical need for balance in the military and non-military dimensions of national security policy. The reality that grand strategy "in war is...necessarily more military than it is in peace" requires "the polity in question...to ensure that...the non-military aspects are not totally neglected." Conversely, the challenge in peacetime is to ensure that "military aspects are not totally neglected (a temptation to which the publics of the post-1919 democracies, recoiling in shock from World War I, were particularly prone)."[55] Clearly, the NDC as a military institution of higher learning primarily concerned with the "great issues affecting Canada's security in an over-populated resource-poor and rapidly changing...unstable world"[56] has failed to heed the latter caution. Although Canadian military officers still have plenty to learn about the sophisticated application of the profession of arms by the time they get to the NDC, they graduate with little knowledge of strategic literature or the various schools of thought associated with national security studies. While presumed to be students of Canadian defence policy, they may not be familiar to any appreciable extent with the works of James Eayrs or Charles Stacey.[57] Surprisingly, the new Emergencies Act is not studied in any detail at the NDC. Neither are students systematically exposed to higher defence planning concepts, including *inter alia* the diplomatic use of naval forces, the compilation of the War Book, national registration, mobilization measures, defence of ports and coastlines, wartime fiscal policy, food controls, and industrial retooling. Precious little structured time is devoted to the study of geography, still the "bones of strategy,"[58] and the use of large-scale maps as opposed to atlases is not encouraged. Lack of student participation in simulated exercises and crisis gaming[59] emphasizing decision-making techniques is equally glaring, and scant attention is paid to examining the relatively abundant scholarship on the determination of national interests. The assertion by some that these are subjects best left to a "war college" as opposed to a "defence college" is not only specious, but reflects a prevalent NDC attitude that as major wars are unlikely to occur in the future, there is little point in even considering the eventuality.[60] To call the NDC the "flagship of the Canadian Forces professional development system" is therefore clearly preposterous.

Paradoxically, neglect of the military sphere has not correspondingly enhanced the status of the NDC as an educational institution. Travel has continued to dominate the curriculum to an unhealthy degree. Basically, trips are booked a year in advance with lectures and study group seminars fitted in around them. While visits to Canada's principal overseas trading partners, the European Community and Japan, can no doubt be justified in terms of national interests, other international junkets are not always undertaken for their strategic or security import. Statements to the effect that "the shopping is better there" or "we should plan to be here on Superbowl Sunday" are often heard during planning conferences not notable for their discussions of substance. It would be foolish, of course, to attempt to deny the obvious educational benefits of travel, but, when cost becomes a major consideration, it is probably worth questioning whether education is *primarily* a function of the simplistic notion "let's fly somewhere and see something." In the grand style of the NDC tour, one luxurious airport inn is pretty much like the next, while the briefness of visits, fatigue, and invariable bouts of stomach upset often exact a formidable toll on the individual and group learning process. The NDC seems to have missed the fundamental point behind Marshall McLuhan's "global village" concept: that "the electric age gave us the means of instant, total field-awareness."[61] The telecommunications revolution, in other words, has brought the world into one's living room. It is, therefore, no longer *absolutely* necessary to travel to gain an understanding of global affairs. Although less exotic, but for far less cost, one can now accompany National Geogaphic teams and the likes of Michael Palin "pole to pole" on television screens or walk round Mayan pyramids on computer videos. Technology, in short, holds the key to bringing the NDC high flyer down to earth.

The learning process for NDC students whether they are flying or not remains remarkably passive. A daily lecture from a "privileged platform" is usually scheduled for 1000 hours and occasionally followed by a second one at 1400 in the afternoon.[62] Although the "privileged platform" is intended to encourage candour by assuring speakers that they will not be quoted by name, either orally or in writing, the effect on free debate is ultimately inhibiting as arguments once made are reduced to unattributable hearsay. To compensate for the deficiencies of set-piece lecture and question periods, students additionally participate in discussion study groups. Unfortunately, these are chaired by inexpert staff colonel "facilitators" chosen for their rank rather than their knowledge. Students are thus essentially left to teach themselves, which approach, though extolled by the NDC, has been questioned by Northrop Frye:

> The etymology of the word "symposium", which means "drinking together", is perhaps of some significance, because in actual symposia, ninety-nine times out of a hundred, one has to be drunk in order to believe what is going on. Nevertheless, the symposium vision in Plato still survives in the mystique of the seminar and the belief that somehow or other a discussion which begins in an unstructured way will eventually achieve structure. Fifty years ago, for example, we have Stephen Leacock saying that if he had an ideal university to found, he would get a room full of students and then go out to hire a few professors when he got around to it. Here again is the belief that the discussion group is the core of education, despite all the evidence proving that what it usually is is a pooling of ignorance.[63]

There is, of course, a place for the seminar in education, but the convenience and relative ease of group solutions should be counter-balanced by serious individual work, reflection, and application. Apart from the requirement to research and write a 20 to 40 page paper on an unfamiliar subject, however, little individual work is actually demanded of NDC students. That ninety per cent of completed papers are generally unacceptable, but accepted anyway, provides a rough indication of the lack of rigour that generally characterizes the course of study.[64] Unfortunately, the NDC unwisely abandoned case study methodology[65], problem-solving approaches, and exercises requiring role playing as complementary means for fostering depth of understanding. The subsequent adoption of an almost entirely "issues" or current events approach contributed to the superficial nature of NDC education, which, for lack of historical perspective, currently tends to treat every topic broached as new.

The drift of the NDC toward academic as well as military irrelevance could no doubt have been arrested by a competent faculty. Unfortunately, this problem has never been seriously addressed. Although Davis attributed the "moderately [un]structured nature" of the course partly to the "intellectual limitations" of the Directing Staff,[66] a later commandant, Major-General L.V. Johnson, maintained that the skills they required were "those of military administration, not of academic knowledge."[67] In a certain sense Johnson was right, for NDC staff directors readily assumed the role of acting as tour guides and organizers as well as inexpert "facilitators." To run a good tour to Africa or Asia became the aim of the game, since foul-ups and *faux pas* could cost one promotion. By 1991-92 it was blatantly obvious that NDC colonels were better qualified to lay on *hors d'oeuvres* and buses than consider questions of grand strategy or national security. Their lack of knowledge of higher military affairs, essentially of their own profession, contrasted starkly with the expertise visiting lecturers usually exhibited in their respective fields. Commandants who tried Voltaire-like to run a corporal's guard by computer calendar[68] and who insisted on supervising captain aides in the operation of their "commandant camcorders" during visits, generally to the exclusion of more important concerns, proved similarly useless pedagogues. Ironically, those best qualified to conduct the NDC course were officers of lower rank but higher academic standing segregated in a separate cell charged with conducting 10-day "mini" NDC courses for other serving officers.[69] Equating rank to knowledge, in short, has been the bane of the NDC. It also remains at the root of Johnson's still pertinent lament that the NDC "doesn't produce graduates with credentials recognized outside the circle of other graduates and a few senior people supporting the College."[70]

Aside from a critical lack of properly qualified faculty, the NDC additionally suffers from a pervading hedonism not necessarily conducive to intellectual pursuits. Disdaining to employ the term "student" to describe course members is but one almost laughable manifestation of this malaise. Another is that not a few candidates continue to arrive looking forward to a rest.[71] Course members travel, moreover, on dedicated Canadian Forces aircraft specially configured for their comfort during long journeys. They are also permitted to consume alcohol and wear relaxed dress on these planes, even though drinking on other service aircraft is strictly forbidden and all passengers are required to dress properly. This waiving of military standards and ethos appears in part to be aimed at impressing civilian students[72], for it is no secret that the NDC lives in constant dread of somehow being found wanting by civil departments of government. Such yearning for approbation may, in turn, explain why during the 1991 northern tour the NDC course travelled by Hercules aircraft from Iqualit to Cape Dorset, ostensibly to sample Inuit culture, but actually to give students an opportunity to purchase quality aboriginal artifacts. In Zimbabwe in 1992 the course Boeing 707 aircraft also made a return trip from Harare to Victoria Falls for the express purpose of sightseeing. The effect of such pandering, of course, is less apt to produce students with a lean and hungry desire for knowledge than sybarites with a thirst for the cocktail circuit. Among commandants and directing staff it is as likely to produce a bureaucratic penchant for high-living and privilege as any real mastery of the profession of arms and all which that entails. For this reason they remain ill-suited to effect worthwhile reform, notwithstanding the fiction that only those who have attended or been part of the NDC can possibly appreciate its value.

On the whole, it is difficult to escape the conclusion that despite a staggering cost the NDC can hardly be considered an institution of higher learning. Behind the facade of its grandiloquent plenums, symposia, theses, and national security analyses, there is woefully little academic or practical depth. For a higher military education, one must go elsewhere; for a higher civilian education, one should not go to the NDC. The defence that the NDC is neither a war college nor a university, but something "very special," is as absurd as the recent suggestion by one commandant that promoting NDC graduates to the highest military ranks was all that was required to confirm the value of what was taught. It would be closer to the truth to say that, bereft of competent faculty, compendium of knowledge, and research capacity, the NDC remains from an educational standpoint neither fish, nor fowl, nor good red herring. True, any fool could not help but benefit from the mind-expanding experience of travel, which may indeed be the point, but in return for money spent one must seriously ask whether a better education could not be attained in other less expensive ways? The answer is undoubtedly yes, for if travel were stricken from the current curriculum, there would be little left to the NDC as an educational institution. The old adage that the NDC is not a college, or national, and has nothing to do with defence, is thus not far off the mark. For what students actually do learn at the NDC, one would hesitate on the basis of this qualification alone to employ them in any capacity, least of all that related to national security. Without drastic reform the NDC will remain what it has been for far too long: an outrageous affront to the hard-pressed Canadian taxpayer who unquestionably deserves a better return on his or her defence dollar.

Lt.-Col. John A. English, 1993

Notes

1. Annex A to 7100-1/91157M (D Cost S) dated 26 September 71.
2. NDC Course Handbook.
3. Letter from Major-General F.J. Norman, Commandant NDC, to Rear Admiral (Retd) J. Aveline, Permanent Secretary of the Conference of Commandants of Defence Colleges of the Atlantic Alliance, 14 October 1988.
4. Sturton Mathwin Davis, "Development and National Characteristics of National Defence Colleges—As a World Phenomenon" (Unpublished Ph.D. Thesis, Queen's University, 1979), iv, 17.
5. David Rutenberg, "Role of a Lifetime," *Dialogue*, 3 (June 1989), 9-11.
6. Don Macnamara, "The Best Year of Your Life," *Dialogue*, 3 (June 1989), 6-8.
7. One commandant, Major-General Scott Clements, was quoted as saying that the NDC aimed "to turn the students into 'better people' in the hope that they will make something of themselves."

Jane Taber, "Opinion," *Ottawa Citizen*, 28 July 1992.

8. Sarah Carnwath, "Canada's Best Kept Secret," *Wardair Revue* (September/October 1989), 43.

9. Opened in 1988 "to enhance public sector management capacities." Jim Armstrong, "Birth of an Institution," *Dialogue*, 3 (June 1989), 12-14.

10. Queen's University Programs for Public Executives 1991. The Queen's program, inaugurated in 1987 as "the only one of its kind in Canada," is conducted for roughly 35 senior executives over a three-week period. See also Charles A. (Sandy) Cotton, "Expanding Horizons for Public Executives, *Dialogue*, 3 (June 1989), 3-5. Queen's additionally conducts a series of three-week Executive Programs for the private sector, which can also be attended by public servants.

11. Don Macnamara, "The Best Year of Your Life," *Dialogue*, 6.

12. Barry D. Hunt, *Sailor-Scholar: Admiral Sir Herbert Richmond 1871-1946* (Waterloo: Wilfrid Laurier University Press, 1982), 149-66; and Brigadier T.I.G. Gray, ed., *The Imperial Defence College and the Royal College of Defence Studies 1927-1977* (London: Her Majesty's Stationery Office, 1977), 1-7, 14, 66-7, 81.

13. Hunt, 1-3, 25, 95, 152, 158-9; and Gray, 5-13, 33-4.

14. James Eayrs, *In Defence of Canada: Peacemaking and Deterrence* (Toronto: University Press, 1972), 61-72; and S. Mathwin Davis, "A Comparative Study of Defence Colleges" (Unpublished M.A. Thesis, Royal Military College of Canada, 1974), 11-12. Selected Canadian senior officers would have continued to attend the IDC and the U.S. National War College, which opened in 1946 and also offered vacancies to Canadians. Eayrs, 70-1.

15. Eayrs, 70-1.

16. Eayrs, 71; and Mary C. O'Connor, "The National Defence College," *Canadian Defence Quarterly*, 2 (Autumn 1971), 41.

17. Gray, 19.

18. The IDC received authorization for a trip to Germany in 1946, but only for 16 students who had not been there. Separate student groups visited North America, the Middle East, and Europe in 1954. The next year a short tour to the British Army of the

Rhine was instituted for all students. A tour to Australia, New Zealand, and Japan was included in 1963, but it was cancelled for reasons of economy in 1976. Overseas tours reached a maximum of seven groups in 1971. A 30-day tour of South America instead of Africa was made in 1966. Gray, 19, 25, 29, 39, 46. In 1988 each RCDS student made one four-week tour abroad. In 1978 the entire NDC class spent 90 days outside of Canada. See Course XXXI "Summary of Useless Information on Field Studies Travel" dated 5 July 1978; and letter from Rear Admiral (Rtd) J. Aveline, Permanent Secretary of the Conference of Commandants of Defence Colleges of the Atlantic Alliance to Major General Norman, Commandant, National Defence College, dated 18 October 1988; and letter Norman to Aveline, 14 October 1988.

19. Colonel George Dramis, "The National Defence College—Does It Satisfy the Canadian Need?" Unpublished paper, National Defence College, 1981, 13; and Davis, "A Comparative Study of Defence Colleges," 10.

20. Adrain Preston, "The Organization of Defence Studies in Canada: A Comparative Analysis," *Brassey's Annual* (London: William Clowes, 1967), 236.

21. Gray, 25-8, 47-51, 76. A former Canadian Army officer and Director of the British Institute for Strategic Studies, Buchan was the second civilian commandant of the IDC. The first was Sir Robert Scott, a diplomat, who served as commandant 1960-1.

22. Memorandum to the Prime Minister from Mr. Gordon Robertson, Clerk of the Privy Council and Secretary to the Cabinet, dated 14 February 1969, covering National Defence College and the Development of Senior Officers in the Public Service, A Report to the Prime Minister, January 1969, under covering letter Frank Milligan to The Right Honourable Pierre Elliott Trudeau, P.C., M.P., dated 6 January 1969; and A Review of Future Developments in Regard to the National Defence College by Brigadier-General S. Mathwin Davis, Acting Commandant Designate-NDC, dated 18 July 1969.

23. Minute by Colonel R.M. Withers, SOPC-2/CDS, 4 March 1969, on letter from E.B. Armstrong, Deputy Minister, to the Chief of

the Defence Staff dated 3 March 1969, covering copy of the Milligan Report.

24. Report of the Officer Development Board, March 1969, 84-5, 249, 265.

25. Ibid., 249-58.

26. Letter, Brigadier-General S. Mathwin Davis, Commandant-Designate NDC, to Brigadier-General D. M. Holman, Commandant NDC, dated 13 August 1969.

27. Ibid., 259-77.

28. Memorandum Major-General J.A. Dextraze, DCPRM, to Chief of Personnel dated 15 April 1969.

29. Memorandum Major-General W.A. Milroy, Chairman CDECPG, to the Chief of the Defence Staff dated 21 March 1969.

30. Davis to Holman 15 August 1969.

31. Davis to Homan 13 August 1969.

32. Sturton Mathwin Davis, "Development and National Characteristics of National Defence Colleges—As a World Phenomenon," 344.

33. Aide Memoire by Davis on Report of Meeting with Mr. Michael Pitfield, Privy Council Secretariat, 16 July 1969; Paper for Presentation Before Cabinet Committee on Priorities and Planning on the Future of the National Defence College; and Outline of Paper to Support NDC Before Cabinet Committee on Policies [sic] and Planning.

34. Memorandum Davis to CDEC-PG, 14 August 1969.

35. Davis, "National Defence Colleges," 344.

36. Dramis, 20-2.

37. Eayrs, 72.

38. Davis, "Comparative Study," 91.

39. Ibid.

40. Hector J. Massey, ed., *The Canadian Military: A Profile* (Toronto: Copp Clark, 1972), 190-1.

41. Course XXXI Yearbook 5 July 1978; and Dramis, 12.

42. Colonel J.D. Harries, "The National Defence College of Canada," *Canadian Defence Quarterly* 2 (Autumn 1989), 41.

43. J.L. Granatstein and Robert Bothwell, *Pirouette: Pierre Trudeau and*

Canadian Foreign Policy (Toronto: University Press, 1990), 234, 260, 378.

44. Teresa Pelton Johnson, "Writing for International Security," *International Security*, 2 (Fall 1991), 172.

45. NDC Course Handbook.

46. Ken Booth, "Security and Emancipation," *Review of International Studies*, 4 (October 1991), 319.

47. Sir John Fortescue opened the third of his Ford Lectures in 1911 with: "War is commonly supposed to be a matter for generals and admirals, in the camp, or at sea. It would be as reasonable to say that a duel is a matter for pistols and swords." Admiral Sir Herbert Richmond, *Statesmen and Sea Power* (Oxford: Clarendon Press, 1947), vii. One is also reminded of the Second Reich's threat to undo the Britsh Empire through "peaceful penetration" by economic means.

48. Lester R. Brown, "Redefining National Security," Worldwatch Paper, 14 October 1977.

49. Thomas L. Dickson, "Redefining National Security: Lexicographical Legerdemain and Confusion," *Defense Analysis*, 1 (1992), 82-3. See also Norman Dixon, *On the Psychology of Military ncompetence* (London: Jonathan Cape, 1976), 189-95 for "anal obsessionality."

50. Booth, 324. Fiscal, social, educational, environmental, and economic security concerns also obtain.

51. C.R. Nixon, "In Defence of Defence," *Ottawa Citizen*, 23 April 1992, A12.

52. To see the Commander of Air Command isolated behind a second-story plexiglass window receiving a morning briefing from his staff through microphones was equally underwhelming.

53. The Seaford House Papers usually comprise up to ten of the best theses written by RCDS course members. They are produced annually in booklet form. Even the Canadian Land Forces Command and Staff College collocated at Fort Frontenac produced a creditable *Quarterly Review*, which contains some of the best 2,500-word essays written by students during a 20-week tactical course.

54. They nonetheless affirmed a process being evolved for the Queen's University School of Business. Rutenberg, 11.

55. Paul Kennedy, ed., *Grand Strategies in War and Peace* (New Haven: Yale University Press, 1991), 169.

56. L.V. Johnson, Notes on Admiral Davis' Proposals for the Development of NDC, December 1981.

57. I refer to James Eayrs' excellent multi-volumed study of national security policy *In Defence of Canada* (Toronto: University Press, 1964-80) and C.P. Stacey's outstanding *Arms, Men and Governments* (Ottawa: Information Canada, 1970).

58. Theodore Ropp, *War in the Modern World* (New York: Collier, 1962) 5.

59. The NDC discontinued the use of crisis management exercises in 1987 supposedly because "mature students didn't wish to play games."

60. A recent commandant went so far as to suggest that such a scenario was a "remote contingency" and therefore not worthy of serious consideration. This was, of course, the same attitude that spawned the notorious "10 Year Rule," which unwisely assumed during the period 1928-1932 that the British Empire would not be engaged in any great war "for the next ten years."

61. Marshall McLuhan, *Understanding Media* (New York: McGraw-Hill, 1964), 47. "As electrically contracted," he wrote, "the globe is no more than a village." This "electric implosion" meant that the "entire world, past and present, now reveals itself to us like a growing plant.... Electric speed is synonymous with light and with the understanding of causes." As he saw it, "In the new electric age of information...the backward countries c[ould] learn from us how to beat us." The problem for developed societies, on the other hand, was to avoid being "numbed" by the information overload. Ibid., 5, 16, 50, 343, 352.

62. In contrast, Queen's Executive Program sessions commence at 0830 hours and continue after lunch until 1530 hours. Evenings between 1930 and 2200 hours are also spent in classrooms. Queen's Executive Program 1993 Brochure.

63. Northrop Frye, *Divisions on a Ground* (Toronto: Anansi, 1982), 126.

64. According to one authority who read many of these papers, roughly 10 per cent were acceptable, 20 per cent could have been improved through proper supervision, and the remainder were not of the standard expected in an institution of higher learning.

65. Which proven methodology continues to be employed by the Queen's Executive Program and the Canadian Centre for Management Development.

66. Davis, "Comparative Study," 95.

67. Johnson, Notes.

68. Voltaire reputedly in his dotage unsuccessfully attempted to make his numerous clocks all run on the same time. Certain military base commanders have also tried to do the same thing with their barrack-room clocks, which for some inexplicable reason never seemed to oblige.

69. This cell of roughly four officers and a secretary was established in 1990 as the Centre for National Security Studies (CNSS). As an integral but junior part of the NDC, the CNSS is forced to abide by the national security parameters set by the larger college. The intensity of its 10-day non-travelling National Security Studies Course is such, however, that its graduates probably emerge having learned almost as much as those of the NDC.

70. Johnson, NDC 4500-0, National Security and the National Defence College, 5 January 1981.

71. Several who opted to leave their families in Ottawa have also connived to obtain ministerial approval for more lavish single accommodation in Kingston beyond that actually warranted.

72. Though not those like the admirable Deputy Minister of Finance, David Dodge, who reportedly insists on walking rather than taking a staff car to work. Deborah Dowling, "Finance's Top Civil Servant Won't Accept Car and Driver," *The Whig Standard*, 25 January 1993.

Index